Computer Games for Learning

Computer Games for Learning

An Evidence-Based Approach

Richard E. Mayer

The MIT Press
Cambridge, Massachusetts
London, England

MIT Press books may be purchased at special quantity discounts for business or sales promotional use. For information, please email special_sales@mitpress.mit.edu.

This book was set in Stone Sans and Stone Serif by Toppan Best-set Premedia Limited.
Printed and bound in the United States of America.

Library of Congress Cataloging-in-Publication Data

Mayer, Richard E., 1947–
Computer games for learning : an evidence-based approach / Richard E. Mayer.
pages cm
Includes bibliographical references and indexes.
ISBN 978-0-262-02757-1 (hardcover : alk. paper)
1. Cognitive learning. 2. Visual learning. 3. Computer games—Psychological aspects. I. Title.
LB1062.M385 2014
371.33'7—dc23

2013046632

10 9 8 7 6 5 4 3 2 1

To Beverly

Contents

Preface

What Is This Book?

Computer games for learning are computer games that are intended to promote learning. The primary goal of this book is to examine what the research evidence has to say about the educational value of computer games for learning, including learning in K–12 education, college education, and workplace training. In short, this book provides a comprehensive, up-to-date, and evidence-based approach to the study of learning with computer games.

Computer Games for Learning distinguishes among three genres of game research: the value-added approach, the cognitive consequences approach, and the media comparison approach (Mayer, 2011). The value-added approach to game research involves comparing the learning outcome measures of students who learn with a base version of a game versus with the base version plus one additional feature. For example, in our research at the University of California, Santa Barbara (UCSB), we have compared students who played the base version of *The Circuit Game* with those who played the same game along with prompts to self-explain each major decision by selecting a reason from a menu (Johnson & Mayer, 2010; Mayer & Johnson, 2010). The base plus self-explanation group performed better than the base group on solving transfer problems on a embedded test, indicating that adding an instructional feature based on self-explanation improved student learning. We also have found that students performed better on solving transfer problems after learning about botany in the *Design-a-Plant* game if an in-game character named Herman-the-Bug communicated by voice rather than through printed text, and if he used conversational as opposed to formal style (Moreno & Mayer, 2000, 2002a, 2002b, 2004; Moreno, Mayer, Spires, & Lester, 2001). In a geology simulation game called the

Profile Game, students performed better on transfer problems if they received training in the names and characteristics of the key concepts before playing the game (Mayer, Mautone, & Prothero, 2002). In an industrial engineering simulation game titled *Virtual Factory*, students performed better on a transfer posttest if an in-game character gave hints and feedback with polite rather than direct wording (Wang et al., 2008). Finally, adding a narrative theme to an adventure game named *Cache 17* intended to teach how mechanical devices work did not improve posttest performance on a transfer test (Adams, Mayer, MacNamara, Koening, & Wainess, 2012). There is a growing research literature based on the value-added approach, so a major goal of the book is to analyze that literature. Overall, value-added research has practical implications for how to design effective educational games as well as theoretical implications for testing cognitive and motivational theories.

The cognitive consequences approach to game research involves comparing the pretest-to-posttest change in learning outcome or cognitive skill (or posttest performance only) for students who played an off-the-shelf computer game for an extended period versus those who did not play the game. For example, in our lab at UCSB we found that students who were required to play the classic video game *Tetris* for five hours showed greater pretest-to-posttest gains in some measures of mental rotation than did students who did not play the game but no differences on most spatial skill measures (Sims & Mayer, 2002). In a different use of the cognitive consequences approach, fifth graders were selected to participate in an after-school club called the Fifth Dimension in which they played a variety of educational computer games over the course of a semester; these students performed better on posttests involving learning a new math game than a matched group of children who were not selected for the club (Mayer et al., 1997; Mayer, Quilici, & Moreno, 1999). There is a substantial literature examining the effects of playing off-the-shelf games on measures of spatial cognition skill (Anderson & Bavelier, 2011), which I also summarize in this book. Overall, the cognitive consequences approach has both practical and theoretical implications concerning people playing video games intended for cognitive improvements.

Finally, the media comparison approach involves comparing the posttest performance on measures of learning outcomes of students who learned by playing a game versus students who learned the same material with conventional media. In our lab at UCSB, for example, we found better transfer test performance from students who learned by playing the *Design-a-Plant* game than from students who learned the same material presented as a series of frames with printed text and graphics (Moreno, Mayer, Spires, & Lester, 2001). In contrast in games with a more complex and intrusive narrative theme, we discovered that students performed worse on a transfer posttest after playing *Crystal Island*, about the spread of disease, or *Cache 17* than from slideshows of the same material featured in each game (Adams, Mayer, MacNamara, Koening, & Wainess, 2012). Although the pitfalls of media research have been well documented (Clark, 2001), this genre of game research has useful practical and theoretical implications for the inclusion of computer games in school settings and beyond.

For the past fifteen years, my colleagues and I at UCSB have been studying the effectiveness of educational games within each of these three research genres, grappling with the key theoretical and methodological issues in game research, and monitoring the growing evidence base on learning with games. I share the fruits of this work in *Computer Games for Learning: An Evidence-Based Approach*—including both our own research program and the full breadth of worldwide research on game effectiveness.

Why Did I Write This Book?

Abt's (1970) *Serious Games*, which is recognized as the first book to consider the potential of games as educational tools, sought evidence for the effectiveness of serious games in education, but of course, there was not a solid research base at the time. Today, many strong claims are made for the educational value of computer games, in the tradition of Abt's unsupported advocacy for serious games. For instance, in an interview on a popular Sunday morning news show on CNN (*Fareed Zakaria GPS*, May 15, 2011), Eric Smith, the CEO of Google, contended, "There's lots of evidence that people learn best in these multiplayer games ... and they learn enormous amounts." In contrast, a recent review of the scientific research on games (Clark, Yates, Early, & Moulton, 2011, p. 269) concluded, "All of the studies that have been published in reputable journals have reached a negative conclusion about learning from games." In another review, O'Neil and Perez (2008) noted, "Computer games were hypothesized to be potentially useful for instructional purposes [but] … there is almost no guidance for game designers and developers on how to design games that facilitate learning" (p. ix). *Games for Learning* helps address this conflict by examining what the research evidence has to say about the effectiveness of computer games intended to promote learning.

What Is in This book?

As can be seen in the table of contents, *Computer Games for Learning* contains eight chapters—three introductory chapters, four chapters analyzing the research base on game effectiveness, and a concluding chapter. Chapter 1, "Introduction: Taking an Evidence-Based Approach to Games for Learning," defines *games for learning,* distinguishes among three research questions about games for learning, provides the rationale for an evidence-based approach, explores the potential educational benefits and drawbacks of games for learning, analyzes the claims of game proponents, and summarizes the history of games for learning.

Chapter 2, "Method: Conducting Scientific Research on Games for Learning," defines and exemplifies three genres of game research (i.e., value added, cognitive consequences, and media comparison) as well as four goals of game research (i.e., what works, when does it work, how does it work, and what happens), applies six principles of scientific research in education to game research (i.e., pose researchable questions, link to theory, use appropriate methods, show coherent reasoning, replicate across studies, and disseminate results), defines and exemplifies three characteristics of experimental research on games (i.e., experimental control, random assignment, and appropriate measures), describes eight common methodological errors in game research, and explores the role of effect size in game research.

Chapter 3, "Theory: Applying Cognitive Science to Games for Learning," examines how the science of learning, science of assessment, and science of instruction are related to learning with games, and shows how basic theories of learning and motivation are relevant to learning with games.

Chapter 4, "Examples of Three Genres of Game Research," provides instances of value-added, cognitive consequences, and media comparison studies that were conducted by our lab at UCSB.

Chapter 5, "Value-Added Approach: Which Features Improve a Game's Effectiveness?," provides a meta-analysis of published studies that compare learning with a base version of a game versus a base version plus one added feature (by reporting means and standard deviations of each group on a measure of learning outcome), and offers a summary of the current state of the field.

Chapter 6, "Cognitive Consequences Approach: What Is Learned from Playing a Game?," supplies a meta-analysis of published studies that compare pretest-to-posttest changes (or posttest scores only) on learning outcome or cognitive skill measures in people who played an off-the-shelf game for an extended time to those changes in people who did not play the game, and provides an analysis of the current state of the field.

Chapter 7, "Media Comparison Approach: Are Games More Effective Than Conventional Media?," offers a meta-analysis of published studies that compare the learning of outcomes of people who learned with a game versus with conventional media, and provides an analysis of the current state of the field.

Chapter 8, "The Future of Research on Games for Learning," summarizes the current state of scientific research on learning with games, offers a research agenda based on promising findings, explores innovations in game research, and suggests domains and contexts in which games have the most potential for fostering learning. The scientific study of learning with games is an emerging field, so this book seeks to help set a productive research course for the future.

The reference list uses an asterisk to highlight each empirical research study contained in the meta-analyses. The book also contains an author and subject index.

Should You Read This Book?

Computer Games for Learning is intended for students and faculty interested in understanding research on educational games; instructional designers and training developers interested in taking an evidence-based approach to educational game design; educational leaders, instructors, and consumers interested in selecting effective educational games; and general readers who are interested in what the research has to say about the value of educational games. It is written for a multidisciplinary audience from a variety of fields including education, psychology, and technology. The book is appropriate for readers of all levels of expertise in the field of educational games ranging from beginners to experts. It could be used in undergraduate and graduate courses concerned with learning and instruction, educational technology, instructional design, and game design.

How Is This Book Different from Other Books on Computer Games?

There are no other books on the market that provide an up-to-date, comprehensive analysis of what the research evidence has to say concerning how to design computer games that promote learning. Three classes of competing books are edited books, visionary books, and practical books, all of which differ from this one in terms of comprehensiveness, educational focus, and/or evidence-based approach.

First, some books on games are edited volumes. There are several fine anthologies concerning research on educational computer games, including Tobias and Fletcher's (2011) *Computer Games and Instruction* and O'Neil and Perez's (2008) *Computer Games and Team and Individual Learning*. Although these books make useful contributions to the field, they are subject to the criticism of most edited books that the approach varies from chapter to chapter and the coverage is selective. Honey and Hilton's (2011) *Learning Science through Computer Games and Simulations* provides a consensus overview of research and development concerning a small segment of the empirical research base by focusing on simulation games for science learning, so it is much narrower than this book. Van Eck's (2010) *Gaming and Cognition* contains chapters that explore conceptual issues and describe development projects in educational games, but does not emphasize empirical research on game effectiveness. Vorderer and Bryant's (2006) *Playing Video Games* as well as Raessens and Goldstein's (2005) *Handbook of Computer Game Studies* offer useful information, but do not spotlight educational issues. In short, *Computer Games for Learning* differs from edited books on computer games by being more coherent and comprehensive in the coverage of research evidence that is relevant to the educational impact of games for learning.

Second, some books on games are visionary volumes, such as Gee's (2007a) *Good Games and Good Learning*, Gee's (2007b) *What Video Games Have to Teach Us about Learning and Literacy*, Shaffer's (2006) *How Computer*

Games Help Children Learn, Squire's (2011) *Video Games and Learning,* Prensky's (2006) *Don't Bother Me, Mom—I'm Learning,* or McGonigal's (2011) *Reality Is Broken.* Such books supply examples of exciting educational game projects and help inspire readers to visualize a future for education in which educational games play a prominent role, but they are not intended to provide comprehensive and convincing evidence. The book you are holding examines the untested claims of game proponents by taking a broad look at the available research base on the effectiveness of educational games.

Third, there are numerous practical guides such as Prensky's (2001) *Digital Game-Based Learning* or Fullerton's (2008) *Game Design Workshop,* but such books are not intended to explore the underlying empirical research base on game design. *Computer Games for Learning* differs from practitioner-oriented books by taking an evidence-based approach.

In short, *Computer Games for Learning* is most similar to Abt's (1970) classic little book *Serious Games,* which attempted to provide an integrated, comprehensive, and evidence-based view of the educational value of games. It can be seen as a modern update that takes an evidence-based approach to examining the educational value of computer games. This book also can be viewed as a complement to *Multimedia Learning* (Mayer, 2009), which takes an evidence-based approach to the instructional design of multimedia presentations.

References

Abt, C. C. (1970). *Serious games.* New York: Viking.

Adams, D. M., Mayer, R. E., MacNamara, A., Koening, A., & Wainess, R. (2012). Narrative games for learning: Testing the discovery and narrative hypothesis. *Journal of Educational Psychology, 104,* 235–249.

Anderson, A. F., & Bavelier, D. (2011). Action game play as a tool to enhance perception, attention, and cognition. In S. Tobias & J. D. Fletcher (Eds.), *Computer games and instruction* (pp. 307–330). Charlotte, NC: Information Age Publishers.

Clark, R. E. (2001). *Learning from media.* Greenwich, CT: Information Age Publishing.

Clark, R. E., Yates, K., Early, S., & Moulton, K. (2011). An analysis of the failure of electronic media and discovery-based learning: Evidence for the performance benefits of guided learning methods. In K. H. Silber & W. R. Foshay (Eds.), *Handbook of improving performance in the workplace* (pp. 263–297). San Francisco: Pfeiffer.

Fullerton, T. (2008). *Game design workshop*. Burlington, MA: Morgan Kaufmann.

Gee, J. P. (2007a). *Good games and good learning*. New York: Peter Lang.

Gee, J. P. (2007b). *What video games have to teach us about learning and literacy* (2nd ed.). New York: Palgrave Macmillan.

Honey, M., & Hilton, M. (Eds.). (2011). *Learning science through computer games and simulations*. Washington, DC: National Academy Press.

Johnson, C. I., & Mayer, R. E. (2010). Adding the self-explanation principle to multimedia learning in a computer-based game-like environment. *Computers in Human Behavior, 26*, 1246–1252.

Mayer, R. E. (2009). *Multimedia learning* (2nd ed.). New York: Cambridge University Press.

Mayer, R. E. (2011). Multimedia learning and games. In S. Tobias & J. D. Fletcher (Eds.), *Computer games and instruction* (pp. 281–306). Amsterdam: Elsevier.

Mayer, R. E., & Johnson, C. I. (2010). Adding instructional features that promote learning in a game-like environment. *Journal of Educational Computing Research, 42*, 241–265.

Mayer, R. E., Mautone, P. D., & Prothero, W. (2002). Pictorial aids for learning by doing in a multimedia geology simulation game. *Journal of Educational Psychology, 94*, 171–185.

Mayer, R. E., Quilici, J. H., & Moreno, R. (1999). What is learned in an after-school computer club? *Journal of Educational Computing Research, 18*, 223–235.

Mayer, R. E., Quilici, J. H., Moreno, R., Duran, R., Woodbridge, S., Simon, R., et al. (1997). Cognitive consequences of participation in a fifth dimension after-school computer club. *Journal of Educational Computing Research, 16*, 353–370.

McGonigal, J. (2011). *Reality is broken*. New York: Penguin Press.

Moreno, R., & Mayer, R. E. (2000). Engaging students in active learning: The case for personalized multimedia messages. *Journal of Educational Psychology, 92*, 724–733.

Moreno, R., & Mayer, R. E. (2002a). Learning science in virtual reality environments: Role of methods and media. *Journal of Educational Psychology, 94*, 598–610.

Moreno, R., & Mayer, R. E. (2002b). Verbal redundancy in multimedia learning: When reading helps listening. *Journal of Educational Psychology, 94*, 156–163.

Moreno, R., & Mayer, R. E. (2004). Personalized messages that promote science learning in virtual environments. *Journal of Educational Psychology, 96*, 165–173.

Moreno, R., Mayer, R. E., Spires, H. A., & Lester, J. (2001). The case for social agency in computer-based teaching: Do students learn more deeply when they interact with animated pedagogical agents? *Cognition and Instruction, 19*, 177–213.

O'Neil, H. F., & Perez, R. S. (Eds.). (2008). *Computer games and team and individual learning*. Amsterdam: Elsevier.

Prensky, M. (2001). *Digital game-based learning*. New York: McGraw-Hill.

Prensky, M. (2006). *Don't bother me, mom—I'm learning*. St. Paul, MN: Paragon House.

Raessens, J., & Goldstein, J. (Eds.). (2005). *Handbook of computer game studies*. Cambridge, MA: MIT Press.

Shaffer, D. W. (2006). *How computer games help children learn*. New York: Palgrave Macmillan.

Sims, V. K., & Mayer, R. E. (2002). Domain specificity of spatial expertise: The case of video game players. *Applied Cognitive Psychology, 16*, 97–115.

Squire, K. (2011). *Video games and learning*. New York: Teachers College Press.

Tobias, S., & Fletcher, J. D. (Eds.). (2011). *Computer games and instruction*. Charlotte, NC: Information Age Publishers.

Van Eck, R. (Ed.). (2010). *Gaming and cognition: Theories and practice from the learning sciences*. Hershey, PA: Information Science Reference.

Vorderer, P., & Bryant, J. (Eds.). (2006). *Playing video games: Motives, responses, and consequences*. Mahwah, NJ: Erlbaum.

Wang, N., Johnson, W. L., Mayer, R. E., Rizzo, P., Shaw, E., & Collins, H. (2008). The politeness effect: Pedagogical agents and learning outcomes. *International Journal of Human-Computer Studies, 66*, 98–112.

Acknowledgments

Computer Games for Learning is the product of fruitful collaborations and discussions with my students at the University of California, Santa Barbara—Deanne Adams, Krista DeLeeuw, Logan Fiorella, Cheryl Johnson, Tricia Mautone, Roxana Moreno, Jill Quilici, and Valerie Sims—and my research collaborators over the years—Rebecca Buschang, Greg Chung, Richard Duran, Lewis Johnson, Deidre Kerr, Alan Koening, James Lester, Harry O'Neil, Bill Prothero, Hiller Spires, Terry Vendlinski, Richard Wainess, and Ning Wang. This book also owes much to my fellow game researchers and scholars, many of who are cited throughout its pages, including Daphne Bavelier, Dexter Fletcher, Shawn Green, Harry O'Neil, Jan Plass, and Sig Tobias. They all have peaked my curiosity and helped sustain my interest in the scientific study of games for learning. In particular, Deanne Adams is the lead author of chapter 6 on cognitive consequences research, and I also appreciate her many useful suggestions concerning this book. I am grateful to work in field that yields so many interesting and important questions.

The Office of Naval Research and National Science Foundation have supported much of my lab's research on games for learning, and this book would not have been written without their aid. I especially appreciate the continuing encouragement and support of Ray Perez. The preparation of the chapters in this book was supported by grants N000141110225 and N000140810018 from the Office of Naval Research.

I wish to thank Phil Laughlin and the talented staff at the MIT Press for their encouragement and support throughout the project.

Finally, the support and warmth of my family means a lot to me, including Ken, David, and Sarah, and their wonderful children, Jacob, Avery, Caleb, James, and Emma. As always, I am immensely grateful to Beverly.

I Introduction

1 Introduction: Taking an Evidence-Based Approach to Games for Learning

Chapter Outline

Summary

Computer games for learning are games and simulations that are intended to promote academic learning. Three important questions about games for learning are:

1. The *value-added question*, Which game features improve learning?
2. The *cognitive consequences question*, What do people learn from playing an off-the-shelf game?
3. The *media comparison question*, Do people learn academic content better from games than from convention media?

An *evidence-based approach* focuses on research data for answering these questions, gleaned from methodologically sound studies and guided by learning theory. Although game proponents make strong claims for the potential of games to revolutionize education, research investigators take a more cautious approach by examining the available scientific evidence. Research on games has a forty-year history, but reviews of game research describe a field that is disorganized and unfocused, with an excess of speculative essays and methodologically flawed studies. This book seeks to systematize research on games for learning by sorting relevant experiments into three research paradigms—value added, cognitive consequences, and media comparison—and then examining published experimental comparisons within these categories that meet the basic requirements of experimental methodology—randomly assigning the participants to groups, matching treatment and control groups on all dimensions except the one being studied, and using appropriate measures of learning outcome.

What Are Games for Learning?

Games for learning are games and simulations that are intended to promote learning. As represented in figure 1.1, playing a game for learning (shown in the box on the left) is intended to cause (indicated by an arrow) a change in the player's knowledge or skill (shown in the box on the right). When games for learning are delivered electronically—such as

Figure 1.1
What are games for learning? Games for learning are games intended to promote learning (i.e., intended to improve performance on measures of learning outcome).

on a desktop computer, laptop computer, tablet, smartphone, or game console—they can be called computer games for learning. The primary goal of this book is to examine what the research evidence has to say about the educational value of computer games for learning. In short, my aim is to provide you with a comprehensive, up-to-date, and evidence-based approach to the study of learning with computers games.

What is a game? Consider the following games: baseball, chess, dominoes, *FarmVille*, *Monopoly*, *Pac-Man*, *Risk*, *Space Invaders*, *Tetris*, and *World of Warcraft*. They share a common set of defining characteristics. A defining characteristic means that to be a game, an educational activity must possess this characteristic. Although you might expect me to engage in a long (and somewhat fruitless) discourse about the definition of games, I have found that it is more useful to list the defining characteristics of games. My list is summarized in table 1.1 (adapted from Mayer & Johnson, 2010): games are interactive, simulated systems that are rule based, responsive, challenging, cumulative, and inviting. This list of features is broad enough to include board games collecting dust in your closet, arcade games you see in the lobby of your local movie theater, console games you like to play on your television screen, and video games you access from the Web to play on your tablet or smartphone.

Let's examine each defining characteristic of games:

Games are rule-based simulated systems The single most important feature of a game is that it represents a simulated system or model based on causal rules that the player can master (in addition to the operational rules of game play, which also must be learned). You are in a simplified world in which things happen for a reason.

Table 1.1
Five defining characteristics of games

Characteristic	Description	Game elements
Rule based	Events occur within a causal system, based on a knowable set of rules	Abstraction of concepts and reality Rules
Responsive	Environment allows for player to act, and responds promptly and saliently	Feedback Replay
Challenging	Environment provides opportunities for success on tasks that are difficult for the player	Goals Levels Conflict, competition, or cooperation
Cumulative	Current state of the environment reflects player's previous actions and allows for assessment of progress toward goals	Time Reward structure
Inviting	Environment is interesting, appealing, and fun for the player	Aesthetics Curve of interest Storytelling

Sources: Characteristics based on Mayer & Johnson (2010); game elements based on Kapp (2012).

Games are responsive Another defining feature is that the simulated system is interactive—that is, it responds promptly and clearly to the player's actions. You can do certain things in the game, and as a result of your actions, something happens in the game.

Games are challenging Another defining feature is that the simulated environment allows for the player to succeed on difficult tasks. You can try to master tasks that are difficult for you.

Games are cumulative Yet another defining feature is that the simulated environment reflects the player's previous actions and allows for assessment of the player's progress toward goals. For example, the game may display your current level or score in an obvious way or the game may be subtle in the way it reflects where you are.

Games are inviting Concerning the impact of games on learners, game scholars have noted that games are fun to play, thereby motivating the learner to keep playing. Thus, a final defining feature is that the simulated

system is interesting or appealing to the player, with the intention of making the experience fun.

These are defining features of games in the sense that to be called a game, a learning environment has to possess these five characteristics.

Kapp (2012) has described *game elements*, which together can turn an educational activity into a game for learning. The right side of table 1.1 shows how some game elements—the building blocks of games—correspond to the five defining characteristics of games for learning. The rule-based characteristic corresponds to the game elements of abstraction (i.e., presenting a simplified model) and rules (i.e., incorporating cause-and-effect rules in the model). The responsive characteristic corrresponds to the game elements of feedback (e.g., telling how far the player is from accomplishing the goal) and replay (i.e., being able to redo an episode or move). The challenging characteristic corresponds to the game elements of goals (e.g., a clear indication of what the player is supposed to accomplish), levels (i.e., a sequence of progressively more challenging tasks), and conflict, competition, or cooperation (e.g., beating an opponent or being part of a team). The cumulative characteristic corresponds to the game features of time (e.g., including a clock) and reward structure (e.g., including scoreboards or awards). Lastly, the inviting characteristic corresponds to the game features of aesthetics (e.g., using appealing graphics), curve of interest (i.e., introducing new aspects of the game in a way to prime interest), and storytelling (i.e., adding a narrative theme).

Kapp (2012) uses the term *gamification* to refer to the process of turning an educational lesson into an educational game by adding appropriate game elements. As Kapp warns, however, simply adding game elements to a lesson does not ensure you will create a good game—that is, a game that people will like to play. To create a good game requires adding appropriate game elements in a way that meets the five defining characteristics in table 1.1. A good game can be seen as a successful art form that is more than the sum of its game elements. Like any successful art form, a good game creates enjoyment, elicits emotional response, provokes thought, and/or motivates action. Although an analysis of the artistic creation of games is beyond the scope of this book, it is worth acknowledging that the larger field of game studies crosses disciplinary boundaries. This book examines how we can mix instructional elements and game elements to create games that promote learning.

What's in a name? You might see a game scholar write about *video games* or *digital games*, whereas a game researcher might use the term *computer games* or *serious games*. In the interests of simplification, I use all of these terms interchangeably. When a game involves a simulated environment in which the player can manipulate things and see what happens, it is sometimes called a *simulation game*. Yet the line between simulation game and computer game is so fuzzy that I include simulations and games in the same category. In short, I use the term *games for learning* to include video games, digital games, computer games, serious games, and simulation games that are intended to promote a change in the learner's knowledge or skill.

When we are interested in moving from games to games for learning, we need to focus on a crucial part of the definition of games for learning: they cause a measurable change in the player's academic knowledge or cognitive skill. A *learning objective* is a clear statement of the to-be-learned knowledge or skill. A *learning outcome* is a clear description of what was learned. Overall, the main goal of most commercially successful computer games is to provide entertainment, but the main goal of games for learning is to foster learning related to academic learning objectives. In short, games for learning are intended to help players build learning outcomes that match the learning objective.

Three Questions about Games for Learning

Our journey into the domain of game effectiveness research starts with some intriguing questions: Do certain features make a computer game more effective in promoting learning? Do people learn useful cognitive skills from playing off-the-shelf games? Do people learn academic content better from playing a computer game than from conventional instruction? If you are interested in answering these kinds of questions, then this book is for you. This book is intended to show you the progress that game researchers have made in addressing these three kinds of questions.

As you can see in table 1.2, three types of questions drive our study of games for learning (Mayer, 2011b):

1. Value-added questions ask how to improve the educational effectiveness of a game. In value-added research, we are interested in whether adding an

Table 1.2
Three kinds of questions about games for learning

Type of question	Example
Value added	Does adding feature X to a game improve learning?
Cognitive consequences	What do people learn from playing an off-the-shelf game?
Media comparison	Do people learn academic content better from games than from conventional media?

Source: Adapted from Mayer (2011b).

instructional feature results in better performance on a test of learning. For example, do students learn better from a game when an on-screen agent speaks to them versus when an on-screen agent communicates the same words through on-screen printed text?

2. Cognitive consequences questions ask about what people learn when they play an off-the-shelf game for an extensive amount of time. In cognitive consequences research, we are interested in whether students who are assigned to play a game show greater improvement on a cognitive skill related to academic performance than students who do not play the game. For example, do students who are asked to play a first-person shooter game for ten hours improve spatial cognition scores more than students who do not play a first-person shooter game?

3. Media comparison questions ask about whether games are a better venue for academic learning than conventional media. In media comparison research, we are interested in whether students learn academic content better from a game or a conventional lesson on the same material. For example, do students who play a physics simulation game perform better on a test of learning as opposed to students who receive the material as a slideshow presentation?

Although each of these questions focuses on a different aspect of the educational effectiveness of games, they all share an important characteristic: *testability*. Testability means it is possible to conduct methodologically sound studies that generate evidence that can address each question. We start our quest for an evidence-based approach to the study of game effectiveness with these three testable questions.

What Is an Evidence-Based Approach?

An evidence-based approach to the study of games for learning involves using appropriate research methods, grounded in learning theory, to yield data needed for answering testable questions. As you can see, this definition has four parts. First, we use appropriate research methods, which are described in chapter 2. Second, we ground research in theories of learning that are relevant to games, as described in chapter 3. Third, we seek to answer testable questions, such as discussed in this chapter. Finally, we gather data, which is at the heart of the scientific study of games for learning. The entire second section of the book, chapters 4 through 7, reviews the research evidence concerning each of our three basic research questions about learning from games.

Why should you care about taking an evidence-based approach to the study of games for learning? In my opinion, an evidence-based approach offers the most helpful way to answer questions about the educational effectiveness of games because it is self-correcting. As research evidence begins to accumulate, we can reject unhelpful accounts of learning with games and construct more useful ones.

In contrast to a scientific approach based on research evidence, there are many other ways to address questions about the educational effectiveness of computer games. Some of the most popular alternatives are to make unsupported claims, point to the popularity of games, or provide anecdotes.

Make unsupported claims Perhaps the most common unscientific approach to addressing questions about the educational effectiveness of computer games is simply to assert that they can be effective. The benefit of this approach is that it frees the author from the pesky requirement of providing a logical argument, based on evidence, and allows for an emotional appeal. The drawback of this approach is that it does not offer an effective way to determine if the claims are justified.

Point to the popularity of games Another popular approach is to highlight the widespread use of video games and the obvious appeal of playing them. The benefit of this approach is that it is based on some evidence, such as the billions of dollars spent on games or millions of game players. The drawback is that this kind of evidence is unhelpful in answering the kinds of basic questions described in the previous section.

Provide anecdotes An additional popular approach is to offer transcripts (or videos) of people talking about the clever things they do while playing games. On the positive side, carefully selected anecdotes provide the appearance of scientific respectability and may even supply interesting examples. On the negative side, there is overwhelming consensus among scientific researchers that reports based primarily on selected anecdotes or testimonials do not qualify as an evidence-based approach.

When you notice that an author resorts to any (or all) of these practices in examining the effectiveness of games for learning, you have left the world of science, and entered the murky realm of speculation and hype.

For the past fifteen years, my colleagues and I at UCSB have been studying the effectiveness of computer games for learning and monitoring the growing evidence base within each of the three research genres listed in table 1.2. We share this research evidence with you in *Computer Games for Learning*—including both our own published research and the full breadth of worldwide published research on game effectiveness.

Potential Benefits and Drawbacks of Games for Learning

Games for learning include both *game features,* intended to motivate learners to engage in game playing, and *instructional features,* intended to foster appropriate cognitive processing during game playing. In particular, the motivating effects of game features include encouraging players to initiate, maintain, and intensely engage in game play. The cognitive effects of instructional features include helping learners attend to relevant material, mentally organize it into a coherent cognitive representation, and integrate it with relevant prior knowledge.

Game features and instructional features in games can affect the three types of cognitive processing during learning shown in table 1.3: *extraneous processing,* which wastes precious cognitive capacity; *essential processing,* aimed at attending to and mentally representing the academic content; and *generative processing,* which involves deeper reflection on the academic content. The goal of effective game design is to reduce extraneous processing, manage essential processing, and foster generative processing (Mayer, 2011a).

Designing effective educational games always involves a tension between including game features for entertainment and instructional

Table 1.3
Three kinds of cognitive processing during game learning

Type of processing	Description	Caused by
Extraneous processing	Cognitive processing that is not related to the instructional objective	Poor game design
Essential processing	Cognitive processing aimed at attending to and mentally representing the academic material	Inherent difficulty of the academic material
Generative processing	Cognitive processing aimed at making sense of the academic material	Player's motivation to engage in learning

Source: Adapted from Mayer (2011a).

Table 1.4
Potential and pitfalls of game features and instructional features in computer games for learning

	Game features	Instructional features
Potential	Game features can promote motivation to learn (increasing generative processing)	Instructional features can promote learning (increasing essential and generative processing)
Pitfalls	Game features can diminish learning (increasing extraneous processing)	Instructional features can diminish motivation to learn (decreasing generative processing)

Source: Adapted from Mayer & Johnson (2010).

features for learning. Table 1.4 summarizes some benefits and drawbacks to game features and instructional features in games for learning. On the positive side, game features can promote the player's motivation, thereby fostering generative cognitive processing and leading to deeper learning outcomes. On the negative side, game features can distract the learner, causing an increase in extraneous cognitive processing and leading to poorer learner outcomes. On the positive side, instructional features can guide the player's attention toward the academic content, thereby supporting essential processing, and prime deeper processing of the academic content, thereby supporting generative processing. On the negative side, though, instructional features can destroy the fun of game playing, thereby reducing the player's interest and diminishing the player's willingness to engage in generative processing.

The challenge of game design is to balance the positive and negative aspects of game features and instructional features to suit the needs of each learner. The research reviewed in this book is intended to help guide this balancing process.

Four Roles in the Field of Games for Learning

As shown in table 1.5, there appear to be four main roles in the area of game scholarship and research:

1. *Visionaries*, who inspire us with a vision of how education could be transformed using games.
2. *Developers*, who dazzle us with a continual stream of exciting new games.
3. *Appliers*, who enrich us by adapting games for learning to use in schools, training programs, and informal situations.
4. *Investigators*, who inform us by conducting scientifically rigorous research about the educational effectiveness of games.

Although visionaries, developers, and appliers tend to focus on the positive potential of games for learning, the lonely gang of investigators (including me) tends to be concerned about the level of scientific support for games.

What Proponents Say: The Claims Are Strong

As shown in table 1.5, visionaries, developers, and appliers usually hold positive views of the educational value of games, so I refer to them as *game proponents*. Their motto might be: "Let's change the educational

Table 1.5
Four roles in game research

Who	What they do	How they view games
Visionaries	Inspire	Positive
Developers	Dazzle	Positive
Appliers	Enrich	Positive
Investigators	Inform	Critical

Table 1.6
What the proponents say: Strong claims for games

"Kids learn more positive, useful things for their future from their video games than they learn in school"
"Good games are problem-solving spaces that create deep learning—learning that is better than what we often see today in our schools"
"The key to solving the current crisis in education will be to use the power of computer and video games to give all children access to experiences, and build interest and understanding"
"Good games lend themselves to systematic understandings"

Sources: Respectively, Prensky (2006, p. 4); Gee (2007, p. 10); Shaffer (2006, p. 67); Squire (2011, p. 36).

world with games." Proponents look for ways to revolutionize education for all learners, and thus seek to bring some much-needed social justice to the world. For example, Table 1.6, gives you a taste of what visionaries have to say about the exciting potential of games to revolutionize education.

As you can see from this set of quotations from leading gaming proponents, there is a tendency to make strong claims. Some of the common claims are: the present educational system has failed; today's students need a different kind of learning experience based on exploration and fun; and games offer an effective alternative educational experience that has been shown to work better than traditional instructional methods. In short, game proponents are telling us that today's schools are not suited to the needs of today's students, but this mismatch can be fixed by incorporating learning experiences based on games and simulations.

Game proponents invite us to imagine a future in which educators agree that students can learn better from well-designed computer games than from conventional instruction in schools. In that future, well-designed games enable students to learn crucial twenty-first-century skills that they will need in their lives. To achieve this vision, the focus will be on pinpointing gaming features that promote learning, thus leading to ever more powerful games for learning. McGonigal (2011), for instance, boldly states: "I foresee games that fix our educational systems" (p. 14).

Within this vision of the future of games in education, I have detected three intertwined themes in the claims that game proponents make:

Media comparison claims Conventional educational practices in schools are not doing a good job at helping students learn, whereas the techniques embodied in well-designed computer games can do a much better job at promoting deep learning.

Cognitive consequences claims Playing computer games can help students develop the kinds of skills and knowledge they will need to lead productive lives in the twenty-first century.

Value-added claims Enhancing certain game features can improve student learning with computer games.

Together, the three intertwined themes share a common premise that well-designed games can promote useful learning.

What Researchers Say: The Evidence Is Weak

If you read popular books by game proponents (as sampled in table 1.6), you would come away feeling enthusiastic about the role of computer games in promoting learning. Game proponents inspire us to consider what games have to offer to education. You might be curious, though, about whether there is convincing evidence to support the claims made for the effectiveness of computer games. Would you be persuaded if the statements about educational games were based on the author's own personal experiences in playing video games or informal observations of someone playing video games? Would you be convinced if the claims were based on anecdotes from teachers and others? How about testimonials from players or game designers? All too often, these are the kinds of evidence supporting the claims being made by game proponents.

What you might not be able to see from the quotes from proponents in table 1.6 is that these strong claims are not always based on strong evidence. Where is the evidence for these claims? Some authors rely on their personal experience or informal observations of children playing games. Some authors offer descriptions of clever games and simulations, sometimes with inspiring words from the developers. Other authors offer what can be called observational *proof-of-concept* studies, which consist of detailed observations of people making discoveries and apparently learning while playing a game.

In contrast, investigators tend to be more critical of the educational value of games, so I refer to them as *game critics*. Their motto might be: "Show me the evidence." Critics keep an open mind and may ultimately

Table 1.7
What game researchers say: Weak evidence for games

"Many strong claims are made for the educational value of computer games, but there is little strong empirical evidence to back up those claims"
"There is relatively little research evidence on the effectiveness of simulations and games for learning"
"There is considerably more enthusiasm for describing the affordances of games and their motivating properties than for conducting research to demonstrate that those affordances are used to attain instructional aims.... This would be a good time to shelve the rhetoric about games and divert those energies to conducting needed research."
"While effectiveness of game environments can be documented in terms of intensity and longevity of engagement ... as well as the commercial success of the games, there is much less solid empirical information about what outcomes are systematically achieved by the use of individual and multiplayer games to train participants in acquiring knowledge and skills. Further, there is almost no guidance for game designers and developers on how to design games that facilitate learning."

Sources: Respectively, Mayer (2011b, p. 281); Honey & Hilton (2011, p. 21); Tobias, Fletcher, Dai, & Wind (2011, p. 206); O'Neil & Perez (2008, p. ix).

turn out to be proponents depending on what the evidence has to say. In this book, I take on the role of an investigator, with the goal of snooping high and low for scientifically rigorous evidence concerning the educational effectiveness of computer games for learning.

For example, table 1.7 gives you a taste of what investigators have to say about the need for empirical evidence grounded in learning theory. A common theme among game investigators is that we need an evidence base of game research. As you can see from the first quote in table 1.7, I fall squarely into the investigator category, and I have written this book as an honest attempt to systematize the available evidence and research-based theory in a way that serves all players in the field of games for learning.

Searching for useful research evidence can be a cumbersome process. Some game scholars occasionally refer to a scientific research study, but when you try to track down the actual paper, you may find that it was not published in a publicly available peer-reviewed research journal. If you are able to locate the paper, you might discover that the research fails to meet the minimal standards of scientific research in education or does not actually test the question under consideration. In a growing number of cases, however, you might be able to successfully track down a reference to a methodologically sound study that does give you new evidence about game effectiveness. These are the cases that form the basis for this book.

Each player in the field of games for learning takes on a specific role or even a combination of roles. The visionaries who inspire us, designers who dazzle us, and appliers who enrich us are not responsible for providing evidence. That job falls on the investigators and that job is the focus of this book.

Historical Overview of Game Research

Abt's (1970) *Serious Games* is widely considered to be the first book to examine the potential of games as educational tools, although, of course, it was published before the video game revolution and before there was much of a research literature on the instructional effectiveness of games. Abt offered the observation that "games are effective teaching and training devices" (p. 13) but he was not able to provide scientifically rigorous evidence to back up his argument. The book you are holding, *Computer Games for Learning: An Evidence-Based Approach,* takes up this challenge by systematically examining the current state of research evidence on games for learning.

Over the years, many strong claims have been made for the educational value of computer games, in the tradition of Abt's unsupported assertion in the foregoing paragraph. In contrast, Hannafin and Vermillion (2008, p. 215) observed: "Games are very motivating and have tremendous potential in education, but despite a rapidly growing base, there is yet insufficient evidence to draw definitive conclusions." Clark, Yates, Early, & Moulton (2011) offer an even harsher analysis of the scientific research on games: "All of the studies that have been published in reputable journals have reached a negative conclusion about learning from games" (p. 269).

Computer Games for Learning helps resolve this conflict between strong claims and weak evidence by providing a systematic and comprehensive analysis of what the research evidence has to say about the effectiveness of games for learning. Should the future of education include a heavy dose of computer games? If so, how should the games be designed and used in education—that is, what makes a good game for learning? Which kinds of learning outcomes, types of learners, and learning contexts are best suited for computer games? All these questions are concerned with the educational effectiveness of computer games—that is, with what works, when it works, and how it works.

The Internet is bursting with sites containing games for learning, and learning games are increasing being included in textbook packages for students. Do these games promote appropriate student learning? There is no shortage of opinions concerning the value of computer games for education. Yet there is a shortage of efforts to back up claims with research evidence. There is no lack of descriptions of computer games and simulations that have been used in educational settings. What is needed, though, is a systematic analysis of research on the educational effectiveness of those games and simulations conducted in a scientifically rigorous way.

To help you get a sense of the work that has been done, let's take a look at eight major reviews of research on the effectiveness of educational computer games over the past twenty years. For each review of game research, table 1.8 lists the author citation, source, number of reviewed studies,

Table 1.8
The big eight reviews of game research

Citation	Source	Number	Method	Description
Randel et al. (1992)	Journal	68	Box score	Effects of games on academic learning (1984–1991)
Hayes (2005)	Report	48	Text	Effects of games on academic learning
Vogel et al. (2006)	Journal	32	Effect size	Effects of games on academic learning
Sitzmann (2011)	Journal	65	Effect size	Effects of simulation games for adult training
Honey & Hilton (2011)	Book	NA	Text	Effects of science games and simulations on learning outcomes
Tobias et al. (2011)	Chapter	95	Text	Findings about games
Connolly et al. (2012)	Journal	70	Text	Effects of games on cognitive skills, learning outcomes, and other measures
Young et al. (2012)	Journal	39	Text	Effects of video games on learning in academic subjects

method for summarizing the results, and a brief description of the focus of the review.

In the journal *Simulation and Gaming*, Randel, Morris, Wetzel, and Whitehill (1992) offered the first major review of game effectiveness. Concerning the media comparison question, of the 68 studies reviewed (including computer-based games and noncomputer-based games), 38 showed no difference in learning from games versus conventional classroom instruction, 27 favored games, and 3 favored conventional instruction. Positive effects for games were strongest in math and language arts, and weakest in social studies. Many methodological problems were identified in the reviewed studies, including questionable controls.

Hayes (2005) produced an unpublished, but highly influential review of 48 game studies for the US Navy. Although Hayes located 274 documents that purported to be on educational games, most could not be used because they only provided the author's opinion on the potential of instructional games (77), did not contain original data (57), did not contain data from a group playing an educational game (56), or had serious methodological flaws (36). Concerning the media comparison question, Hayes found evidence that games are effective for some learning tasks, but they are not superior to other instructional approaches. In regard to the value-added question, Hayes found evidence that games are more effective if they include instructional support, feedback, and supplemental instruction aimed at helping the learner interpret the game experience in terms of instructional objectives. Hayes noted that the game research literature is "filled with ill-defined terms and plagued with methodological flaws" (p. 6).

The first major meta-analysis of the game effectiveness literature was published in the *Journal of Educational Computing Research* by Vogel, Vogel, Cannon-Bowers, Bowers, Muse, and Wright (2006). Although their search located 248 studies on educational games, only 32 actually met their minimal standards for inclusion. Concerning the media comparison question, "significantly higher cognitive gains were observed in students utilizing interactive games or simulations versus traditional teaching methods" (Vogel et al., 2006, p. 233). Concerning the value-added question, it appears that learning gains were equivalent with low-realism graphics or high-realism graphics. The authors note that their review was hindered by the fact that too many studies had to be excluded due to methodological flaws, such as the lack of a control group or statistical data.

Sitzmann (2011) published a more restricted meta-analysis in *Personnel Psychology* focusing on the effectiveness of computer-based simulation games for adult training. Of the 264 identified studies, only 40 met the minimal criteria for inclusion, but these along with 15 additional reports found in literature reviews yielded a total of 65 experimental comparisons. Concerning the media comparison question, the game group outperformed the conventional group on posttest measures of declarative knowledge (d = 0.28 based on 39 comparisons), procedural knowledge (d = 0.37 based on 22 comparisons), and delayed declarative knowledge (d = 0.22 based on 8 comparisons). As for the value-added question, the positive effects of simulation games were strongest when the game was used as a supplement to other instruction, learners had unlimited access to the game, and the game involved learner interactivity. Finally, Sitzmann found a huge *file-drawer effect* in which games were found to be superior in published papers (d = 0.52 based on 44 comparison), but not in unpublished papers (d = –0.10 based on 16 comparisons).

Honey and Hilton (2011) edited a consensus report commissioned by the National Research Council on the effectiveness of computer games and simulations in science education. The report examined media comparison, value-added, and cognitive consequences studies involving many well-known science simulation games, yet Honey and Hilton were forced to conclude: "Evidence for the effectiveness for supporting science learning is emerging, but is currently inconclusive. To date the research base is very limited" (p. 54). The report observed "many gaps and weaknesses in the body of research on the use of simulations and games for science learning" (p. 55), and found "most studies lack control groups, making it difficult to conclude that the game or simulation caused any learning gains" (p. 21).

Also in 2011, Tobias, Fletcher, Dai, and Wind (2011) published a wide-ranging review of 95 educational game studies in a chapter in an edited book by Tobias and Fletcher (2011) titled *Computer Games and Instruction*. The review included examples of media comparison, value-added, and cognitive consequences research as well as some observational studies. It summarized some potentially positive findings showing that games can be effective learning venues, adding certain features can make them more effective, and cognitive processing can be improved by playing certain off-the-shelf games. The authors called for less "rhetoric about games" and investing more energy in "conducting needed research" (p. 206).

Connolly, Boyle, MacArthur, Hainey, and Boyle (2012) published another wide-ranging review of 70 educational game studies in *Computers & Education.* Of 7,392 papers identified in their searches, only 70 met their minimal criteria for inclusion. The review described instances of high-quality studies using media comparison, value-added, and cognitive consequences approaches to game research, including some with positive findings for learning academic content and cognitive processing skills. The authors noted that many published research papers had to be excluded on methodological grounds.

The last entry in table 1.8 summarizes a 2012 review in *Review of Educational Research* by Young, Slota, Cutter, Jalette, Mullin, Lai, Simeoni, Tran, and Yukhymenko (2012) that examined the effects of video games on K–12 academic learning. Of the 363 identified studies, only 39 met the minimal criteria for inclusion in the review, including some in science, language learning, math, physical education, and history. Concerning the media comparison approach, the authors concluded that the research provided some support for the positive effects of video games on language learning, history, and physical education, but "little support for the academic value of video games in science and math" (p. 61). They called for research that identifies how best to implement games within classroom and social environments.

As you can see, the reviews listed in table 1.8 (as well as other reviews, such as Anderson & Bavelier, 2011; Ke, 2009; Lee, 1999) show that the research literature on educational games is highly diverse, disorganized, and unfocused, with an unusually high number of methodologically flawed studies. In this book, I attempt to systematize the literature by focusing on three basic research paradigms: value-added research, cognitive consequences research, and media comparison research (as described more fully in chapter 2). By sorting experimental studies into these three categories, my goal is to bring more order and organization to a field that seems unfocused. I strive to ensure methodological rigor in this book by only examining experimental comparisons within each of these paradigms that meet the scientific requirements of random assignment, experimental control, and appropriate measures, as also described in chapter 2. By clearly specifying the conditions of an acceptable experimental comparison, my goal is to elevate the status of scientific methodology in a field that needs high-quality research. These more organized reviews of rigorous game research are presented in the evidence section of this book (i.e., chapters 4, 5, 6, and 7). The next chapter (i.e., chapter 2) explores the methodological basis for game research.

References

Abt, C. C. (1970). *Serious games*. New York: Viking.

Anderson, A. F., & Bavelier, D. (2011). Action game play as a tool to enhance perception, attention, and cognition. In S. Tobias & J. D. Fletcher (Eds.), *Computer games and instruction* (pp. 282–306). Charlotte, NC: Information Age Publishing.

Clark, R. E., Yates, K., Early, S., & Moulton, K. (2011). An analysis of the failure of electronic media and discovery-based learning: Evidence for the performance benefits of guided learning methods. In K. H. Silber & W. R. Foshay (Eds.), *Handbook of improving performance in the workplace* (pp. 263–297). San Francisco: Pfeiffer.

Connolly, T. M., Boyle, E. A., MacArthur, E., Hainey, T., & Boyle, J. M. (2012). A systematic review of empirical evidence on computer games and serious games. *Computers & Education, 59*, 661–686.

Gee, J. P. (2007). *Good video games and good learning*. New York: Peter Lang.

Hannafin, R. D., & Vermillion, J. R. (2008). Technology in the classroom. In T. L. Good (Ed.), *Twenty-first-century education: A reference handbook* (Vol. 2, pp. 209–218). Thousand Oaks, CA: Sage.

Hayes, R. T. (2005). *The effectiveness of instructional games: A literature review and discussion*. Naval Air Warfare Center Training Systems Division, Technical Report 2005–004, Orlando, FL.

Honey, M., & Hilton, M. (Eds.). (2011). *Learning science through computer games and simulations*. Washington, DC: National Academy Press.

Kapp, K. M. (2012). *The gamification of learning and instruction*. San Francisco: Pfeiffer.

Ke, F. (2009). A qualitative meta-analysis of computer games as learning tools. In R. E. Ferdig (Ed.), *Effective electronic gaming in education* (Vol. 1, pp. 1–32). Hershey, PA: Information Science Reference.

Lee, J. (1999). Effectiveness of computer-based instructional simulation: A meta-analysis. *International Journal of Instructional Media, 26*(1), 71–85.

Mayer, R. E. (2011a). *Applying the science of learning*. Upper Saddle River, NJ: Pearson.

Mayer, R. E. (2011b). Multimedia learning and games. In S. Tobias & J. D. Fletcher (Eds.), *Computer games and instruction* (pp. 281–305). Greenwich, CT: Information Age Publishing.

Mayer, R. E., & Johnson, C. I. (2010). Adding instructional features that promote learning in a game-like environment. *Journal of Educational Computing Research, 42*, 241–265.

McGonigal, J. (2011). *Reality is broken: How games make us better and they can change the world*. New York: Penguin Press.

O'Neil, H. F., & Perez, R. S. (Eds.). (2008). *Computer games and team and individual learning*. Amsterdam: Elsevier.

Prensky, M. (2006). *Don't bother me, mom—I'm learning*. Saint Paul, MN: Paragon Press.

Randel, J. M., Morris, B. A., Wetzel, C. D., & Whitehill, B. V. (1992). The effectiveness of games for educational purposes: A review of recent research. *Simulation and Games, 23*, 261–276.

Shaffer, D. W. (2006). *How computer games help children learn*. New York: Palgrave Macmillan.

Sitzmann, T. (2011). A meta-analytic examination of the instructional effectiveness of computer-based simulation games. *Personnel Psychology, 64*, 489–528.

Squire, K. (2011). *Video games and learning*. New York: Teachers College Press.

Tobias, S., & Fletcher, J. D. (Eds.). (2011). *Computer games and instruction*. Charlotte, NC: Information Age Publishers.

Tobias, S., Fletcher, J. D., Dai, D. Y., & Wind, A. P. (2011). Review of research on computer games. In S. Tobias & J. D. Fletcher (Eds.), *Computer games and instruction* (pp. 525–545). Charlotte, NC: Information Age Publishing.

Vogel, J. J., Vogel, D. S., Cannon-Bowers, J., Bowers, C. A., Muse, K., & Wright, M. (2006). Computer gaming and interactive simulations for learning: A meta-analysis. *Journal of Educational Computing Research, 34*, 229–243.

Young, M. F., Slota, S., Cutter, A. B., Jalette, G., Mullin, G., Lai, B., et al. (2012). Our princess is in another castle: A review of trends in serious gaming for education. *Review of Educational Research, 82*, 61–89.

2 Method: Conducting Scientific Research on Games for Learning

Chapter Outline

Summary

What is scientific research on game effectiveness? This chapter examines scientifically rigorous research methods for assessing the effectiveness of computer games intended to improve academic learning. It presents three experimental designs for conducting controlled experiments on computer games for learning (i.e., value-added, cognitive consequences, and media comparison designs), and shows how they are related to four research goals (i.e., determining what works, when does it work, how does it work, and what happened). Moreover, it shows how six principles for conducting scientific research in education are related to game research. It then summarizes three characteristics of experiments (i.e., experimental control, random assignment, and appropriate measures), and relates them to eight ways to conduct a useless study on game research. The chapter concludes by exploring the central role of effect size in game research.

Three Designs for Research on Game Effectiveness

Table 2.1 summarizes the experimental design for each of the three genres of game research described in chapter 1—value-added, cognitive consequences, and media comparison experiments (Mayer, 2011b). As you can see, each experimental design involves a comparison between a treatment group and a control group. All three kinds of experimental designs adhere to the basic requirements of random assignment (i.e., the participants are randomly assigned to treatment and control groups), experimental control (i.e., the treatment and control groups are equivalent on all relevant dimensions except for the one being studied), and appropriate measures (i.e., learners are tested on an academically useful learning outcome or cognitive skill) as discussed later in this chapter. The three kinds of experimental designs, however, differ with respect to the question being addressed as well as the characteristics of the treatment and control groups.

Value-Added Experiments

Suppose you wanted to know if you should add on-screen characters to a physics game that clap every time the player makes a correct move. In

Table 2.1
Three experimental designs for research on games for learning

Type of research	Research focus	Example
Value added	To determine the effectiveness of adding a new feature to a game	Does adding an on-screen character who gives feedback improve learning in a physics game?
Cognitive consequences	To determine the degree to which playing an off-the shelf game causes learning in the player	Does playing an action game cause improvements in spatial and perceptual attention skills?
Media comparison	To determine whether a game is more effective than a conventional medium	Do people learn more about fractions from playing a game than from filling out worksheets on the same content?

short, you want to know if adding feature X to a game will cause players to learn better. This type of question is best addressed by a value-added research design, which has the following basic characteristics for the control and treatment groups:

Control group: People play a base version of the game.

Treatment group: People play the same game with one feature added.

As you can see, the goal of value-added experiments is to be able to determine whether adding a particular feature to a game causes a useful change in the learner's academic knowledge.

Cognitive Consequences Experiments

Suppose you wanted to know if playing a particular action game for an hour every day over two weeks would increase people's spatial and perceptual attention skills. In short, you want to know if playing game Y will cause players to improve their cognitive skills related to academic learning. This type of question is best addressed by a cognitive consequences research design, which has the following basic characteristics for the control and treatment groups:

Treatment group: People play an off-the shelf game for an extended period.

Control group: People engage in some unrelated computer-based activity for an extended period.

As you can see, the goal of cognitive consequences experiments is to determine whether playing a particular game causes an educationally useful change in the player's cognitive skills.

Media Comparison Experiments

Suppose you wanted to know if students learn fractions better from playing a math game on tablet computers in their classroom for twenty minutes a day for ten days or from completing worksheets that cover the same problems as the game during the same period. This type of question is best addressed by a media comparison research design, which has the following basic characteristics for the treatment and control groups:

Treatment group: People learn academic content by playing a game.

Control group: People learn the same content with conventional media.

The goal of media comparison experiments is to determine whether people learn academic content better from games or conventional media.

Four Goals of Game Research

In general, most studies of games address one (or more) of four major goals—aimed at determining what works, when it works, how it works, and what happens (Mayer, 2011a). Table 2.2 summarizes examples and research methods for each of these goals. The three experimental designs listed in table 2.1 are mainly aimed at determining what works, and can be adapted to help determine when and how it works. Furthermore, their construction can be informed by determining what happens.

What Works?

The key question in game effectiveness research is "Does it work?" as summarized in the top portion of table 2.2. Each of the research approaches listed in table 2.1 is aimed at the what-works goal: determining the learning effects of adding an instructional feature to a game versus having people learn from the base version of the game (i.e., value-added approach), the cognitive effects of having people play a game versus not play a game (cognitive consequences approach), and the learning effects of playing a game as opposed to learning in a nongame environment (i.e., media comparison approach). In each case, the aim is to determine what works in games for learning.

Table 2.2
Four goals of game research

Goal	Example	Method
What works?	Value added: Does adding feature X to a game improve learning? Cognitive consequences: Does playing game Y improve cognitive skills? Media comparison: Do people learn better from games than from conventional media?	Experimental comparison
When does it work?	Value added: Are the effects of feature X stronger for certain kinds of learners? Cognitive consequences: Does playing game Y improve cognitive skills better for certain kinds of learners? Media compassion: Do certain kinds of people learn better from games than from conventional media?	Factorial comparison
How does it work?	Value added: Why does adding feature X to a game improve learning? Cognitive consequences: Why does playing game Y improve cognitive skills? Media comparison: Why do people learn better from games than from conventional media?	Experimental comparison, interview, observation, survey
What happens?	What do people do when they play a game?	Observation, survey, interview

Determining whether a game or game feature promotes the intended learning change is a fundamental goal in game effectiveness research. The most appropriate research method for assessing the instructional effects of an intervention is an experimental comparison between the learning outcome performance of a treatment group and a control group (Mayer, 2009; Shavelson & Towne, 2002).

When Does It Work?

A secondary issue in game effectiveness research is to determine the conditions under which an effect is strong and the conditions under which an effect is weak. Thus, the second section of table 2.2 takes the what-works goal one step further by asking if the effect is equivalent for different subgroups, different kinds of instructional contexts, and different kinds of instructional content. The most appropriate research method for assessing the generality of instructional effects of games and game features is a quasi-experimental comparison in which the differences between treatment and control groups are examined for various subgroups, learning contexts, or types of content.

How Does It Work?

The next step in game effectiveness research is to explain the effects within the context of a theoretical model. The third section of table 2.2 focuses on the goal of determining how a game or game feature causes a change in the learner. For example, if playing a simulation game with competition works better than playing the game without competition, we want to know the mechanism by which competition improves learning. In this case, perhaps the competition causes the study participant to be more motivated, so it would be worthwhile to include measures of the learner's level of motivation in the study. The most appropriate research method for how-does-it-work goals involves experiments with some form of observation—including direct observation of the participant's behavior or physical state, or self-report from the participant about internal states and gratifications from game play in response to either an interview or questionnaire. Data about physical states can also be collected by sensors, which are increasingly used as inputs to games to gather data about players' level of anxiety; physiological measures such as heart rate and blood pressure; and body movements such as gestures and the number of steps taken during a certain period. Eye-movement and brain activity measures can also be used. Moreover, games often store data about players' patterns of use and their progress in the game—such as how much time each player spends playing the game per game play session, which components of the game they choose to play if choice is offered, how frequently they make errors and need help, what kinds of errors they usually make, and how often they succeed and at which game challenges. These data can be compared with data collected through observations, self-reports, and sensor readings to develop a fuller picture of the study participants along with how they are responding to specific features and events within a game.

What Happens?

The fourth section in table 2.2 shows one of the most prevalent forms of game research—describing what the participants do when they play a game. The source of data may be a log file that records every action by the participant, a video of the screen as the participant plays the game in a research laboratory, a recording of the discussion between two participants as they jointly play a game, a retrospective interview in which the participant is asked questions about game play, a thinking-aloud protocol in which participants describe their thoughts continuously during game play, or a survey in which a participant answers questions or makes ratings either on the screen or paper. The main goal is to describe something about game playing, such as the individual's evaluation of the game, level of fatigue or frustration, level of enjoyment, or number of thoughts generated about a persuasive message or game experience in order for researchers to capture how deeply the study participant is processing the information.

On the positive side, answering the question "What happens?" can generate useful preliminary information that can be used to suggest more focused hypothesis-driven research. Descriptive research can add a useful level of richness to our understanding of game players and the environments in which they play. This research methodology can easily be adapted and used in randomized experiments that contrast two slightly different versions of a game as a way to test theories about players' psychological processing of games as well as the gratifications they experience with various game types.

On the negative side, collecting observational data for data's sake can be a fruitless enterprise, and making sense of the collected data can be an insurmountable challenge. In some cases, descriptive studies take the form of anecdotes, which generally lack the scientific rigor needed to make supportable conclusions. In other cases, they take the form of quantitative depictions of highly unrepresentative or extremely small samples, with little potential for generality. In short, it may be easy to collect observational data, but the research challenge is to use the data to make a significant contribution to the field. In this book, I examine the first three questions listed in table 2.2, with a primary focus on the first question because it is best addressed by the field's current state of development.

Six Principles of Scientific Research in Education

This book is based on the idea that evidence from scientifically rigorous research has a useful role to play in helping to design games for learning. Before examining some of that research in the next section, let's begin by considering how to conduct scientific research on game effectiveness. An appropriate starting point is to ask, "What exactly do you mean by scientific research?"

In a landmark report from the National Research Council, titled *Scientific Research in Education*, Shavelson and Towne (2002, p. 3–4) propose six consensus principles of educational research that are relevant to conducting research on game effectiveness, as shown in table 2.3.

Pose Significant Questions That Can Be Investigated Empirically

Many claims are made about games, but to be a scientific hypothesis, the claim must be stated in a way that allows for it to be tested empirically. Shavelson and Towne (2002) clearly make this point: "Testability and refutability of scientific claims or hypotheses is an important feature of scientific investigations that is not typical in other forms of inquiry" (p. 3). The first step in game research is to develop a clearly testable hypothesis, so in this book I focus on what I call three genres of game research (Mayer, 2011b), as outlined in chapter 1:

- Value-added research: Does adding feature X to a game improve student learning?
- Cognitive consequences research: Does playing game Y improve cognitive skills needed for academic learning?

Table 2.3
Six principles of scientific research in education

1. Pose significant questions that can be investigated empirically
2. Link research to relevant theory
3. Use methods that permit direct investigation of the question
4. Provide a coherent and explicit chain of reasoning
5. Replicate and generalize across studies
6. Disclose research to encourage professional scrutiny and critique

Source: Shavelson & Towne (2002, p. 3–5).

- Media comparison research: Does playing a game result in better academic learning than receiving the same content through conventional media?

The goal in formulating these three research questions—the centerpieces of this book—is to satisfy scientific principle 1, because each is intended to pose a significant question about game effectiveness that can be investigated empirically.

Link Research to Relevant Theory

Shavelson and Towne (2002) note: "Every scientific inquiry is linked ... to some overarching theory or conceptual framework that guides the entire investigation" (p. 3). Chapter 3 outlines the conceptual frameworks that guide this investigation of game effectiveness—including cognitive theories of how people learn and motivational theories of how people exert effort to learn.

Use Methods That Permit Direct Investigation of the Question

Shavelson and Towne (2002) state: "Methods can only be judged in terms of their appropriateness and effectiveness in addressing a particular research question" (p. 6). In short, "the method used to conduct scientific research must fit the question posed" (p. 63). In a subsequent section of this chapter, I describe experimental methods for answering several types of research questions concerning game effectiveness. There is consensus that experimental methods are most appropriate for addressing questions about instructional effectiveness, such as the value-added, cognitive consequences, and media comparison questions listed previously (Mosteller & Boruch, 2002; Phye, Robinson, & Levin, 2005). Shavelson and Towne (2002) point to the need for randomized controlled experiments to answer questions like our three questions about game effectiveness: "From a scientific perspective, randomized trials (we also use the term *experiment* to refer to causal studies that feature random assignment) are ideal for establishing whether one or more factors caused a change in an outcome" (p. 110). I refer to this approach as making *experimental comparisons* for each of the three game research questions addressed in this book.

Provide a Coherent and Explicit Chain of Reasoning

Shavelson and Towne (2002) underscore the necessity of drawing conclusions based on the available evidence: "At the core of science is inferential reasoning" (p. 4). Within each research chapter (chapters 4 through 7), I draw conclusions about game effectiveness based on patterns in the available data, and in final chapter, I suggest some promising future directions.

Replicate and Generalize across Studies

Replication is at the heart of the scientific study of game effectiveness and forms the basis for the meta-analyses reported in chapters 5 through 7. Shavelson and Towne (2002) summarize the scientific principle of replication as follows: "Ultimately, scientific knowledge advances when findings are reproduced in a range of times and places and when findings are integrated and synthesized" (p. 4). The meta-analyses reported in chapters 5 through 7 are intended to serve the scientific principle of replication and generalization across studies.

Disclose Research to Encourage Professional Scrutiny and Critique

Shavelson and Towne (2002) state: "Scientific studies do not contribute to a larger body of knowledge until they are widely disseminated and subjected to professional scrutiny by peers" (p. 6). An important goal of this book is to highlight and scrutinize the small but growing research base on the effectiveness of games for learning, such as reported in chapters 4 through 7.

Three Characteristics of Experimental Research on Game Effectiveness

There is consensus among educational researchers that the research method should match the research goal (Shavelson & Towne, 2002). When the goal is to determine the instructional effectiveness of games or game elements (e.g., what works or when does it work), experimental research is indispensable because it is designed to determine whether a particular treatment causes a particular learning outcome (Mayer, 2009). Table 2.4 defines and exemplifies the three central characteristics of an experiment: experimental control, random assignment, and appropriate measures (based on Mayer, 2011a). The scientific rigor of game effectiveness research depends on the presence of these three elements in experimental comparisons.

Table 2.4
Three characteristics of experiments

Characteristic	Definition	Example
Experimental control	Treatment and control groups receive identical treatments except for one element	One group plays a game and another group plays the same game with a scoreboard added to foster competition
Random assignment	The players are randomly assigned to the treatment and control groups	For 100 players, 50 are randomly selected to be in the treatment group and 50 are randomly selected to be in the control group
Appropriate measures	Mean (*M*), standard deviation (*SD*), and sample size (*n*) are reported for each dependent measure of learning outcome	The scores on a posttest are *M* = 5.0, *SD* = 1.0, and *n* = 50 for the treatment group, and *M* = 4.0, *SD* = 1.0, and *n* = 50 for the control group

Experimental Control

Experimental control refers to the idea that the treatment group and control group (sometimes called the comparison group) receive the same treatment except for the one element that is being manipulated. In game research, the value-added approach offers the opportunity for the best experimental control. The control group receives the base version of the game, whereas the treatment group receives the same game with a feature added to it. When implemented properly, the only difference between the two games is the added feature (e.g., the presence or absence of an on-screen character who claps every time that the player makes a correct move).

The cognitive consequences approach has the potential to provide the second-best venue for experimental control. The treatment group plays an off-the-shelf game for an extended period, whereas the control group plays an unrelated game or engages in an unrelated activity for the same period, so the only difference between the groups is the content or format of the game being played or the unrelated activity (e.g., a first-person shooter game that requires spatial skills versus a quiz game that does not). In some cases, the cognitive consequences approach is implemented by having an experimental group that plays a game and control group that does not.

A problem with the game versus no-game comparison is that there are many differences between playing a game and not playing any game at all, so it is not possible to pinpoint exactly which aspects of game content or game playing caused an effect on learning or behavior. For this reason, cognitive consequences research can be followed up by value-added (or value-subtracted) research that compares the versions of the same game.

Finally, the media comparison approach runs the risk of affording the least amount of experimental control of any of the three genres of game research. The treatment group plays the game, whereas the control group receives the same information in another format, so the only difference between the groups appears to be the delivery medium. In converting the content from the game to another format such as a face-to-face lecture or an online slideshow, however, you may be creating many differences between the groups, including how the material is segmented, how much extraneous material is presented, the voice of the instructor, and so on. Also, games by their very nature are nonlinear. Players can choose what to do and where to go in a game, and so they are exposed to different content in unplanned ways, and they also see different content depending on whether they fail or succeed. Many other forms of media are completely linear, such as a slideshow presentation or video, so the instruction and information appear in a set order for a set amount of time. The results of media comparison studies must be interpreted in light of possible threats to experimental control, yet these threats also should be balanced against the practical usefulness of comparing two commonly used delivery modes.

Random Assignment

Random assignment requires that people must be selected for each group based on chance. If you simply compare game players to nonplayers, for example, you have violated the requirement for random assignment because people selected themselves to be in the game player and nonplayer groups. If you find differences between the two groups on an important learning outcome measure, you do not know whether it can be attributed to game playing, or preexisting differences in the kinds of people who choose to play or not play. Random assignment to treatment and control groups is a hallmark of experimental comparisons for each of the three experimental designs for game research.

Appropriate Measures

In the empirical study of games for learning, we are interested in using games to foster educationally useful changes in what people know and can do, so it is crucial to have relevant and valid measures of learning outcome, including means, standard deviations, and sample size. In short, we want measures that tell us what was learned. For example, simply asking people to rate through a self-report survey how much they learned as a result of playing a game is usually not a particularly useful or accurate outcome measure. An exception is when, for instance, perceived learning can serve as a mediating variable that is being tested to see if it is associated with stronger or weaker learning outcomes. Similarly, it can be useful to determine the relation between learning outcome measures and theoretically important physiological or survey scale measures. If the goal is to improve understanding of how a disease system works, then a problem-solving transfer test is appropriate. If the goal is to improve arithmetic with fractions, then a fraction computation test is appropriate. If the goal is to improve spatial ability, then a mental rotation test is appropriate. Although the measures are quantitative—that is, they are represented as a number—they can be followed up by qualitative measures—such as verbal descriptions—to add richness and answer questions about why it works.

Eight Ways to Conduct a Useless Study on Game Effectiveness

If your goal is to consult the evidence on game effectiveness, you will find that there is no shortage of useless studies. For example, O'Neil, Wainess, and Baker (2005) found over four thousand published studies on games in a search of the literature, but only nineteen met the minimal requirements of reporting an experimental comparison using a measure of learning outcome. Based on my own search through the game literature in which the vast majority of papers do not meet the simple criteria listed in table 2.4, I offer my eight favorite ways to conduct a useless study on game effectiveness in table 2.5.

Use the Wrong Dependent Measure

In a research report, people rate their enjoyment higher when they play an educational game on a tablet rather than on a desktop computer. In another research report, people reach a higher heart rate when playing a 3-D version

Table 2.5
Eight ways to conduct a useless study on game effectiveness

Description example
1. Use the wrong dependent measure
2. Don't bother to include a control group
3. Compare apples to oranges
4. Use preexisting groups
5. Skimp on the number of participants
6. Rely on anecdotes
7. Use an ineffective game
8. Do not report on means, standard deviations, and sample size

of an educational game than in a 2-D version. In yet another research report, groups of students have more far-ranging discussions while playing an adventure game than while viewing a narrated slideshow covering the same material. What's wrong with these studies as sources of evidence about instructional effectiveness? In each case, the dependent measure involves something other than academic learning outcome—that is, the dependent measure fails to tell us what was learned. In short, they violate the requirement to have *appropriate measures*.

A crucial step in game effectiveness research concerns deciding what to measure. Some researchers focus on player ratings of either how much they liked the game or how motivated they are to play it. Some researchers ask players to rate the effectiveness of the game. Others diligently measure every key that the player presses. Some carefully observe what the players do and say as they plan, and then compile lengthy transcripts from which they select small fragments for publication. Some researchers focus on game sales as an indication of game effectiveness. Others who have access to physiological instrumentation may measure heart rate or pupil dilation during game play, or even which areas of the brain are activated.

What is wrong with these measures? Each one can provide some useful information, but none tell us what the player has learned. The single most important dependent measure in game effectiveness research is academic learning outcome. In my experience, the most aggravating problem in game effectiveness research occurs when researchers measure anything and everything other than the one thing we care most about—academic learning outcome. The gold standard in game effectiveness research is to measure a change in what the learner knows or can do.

Consider an experimental study comparing a game group and a control group in which both groups score near 100 percent on a learning outcome test, or both groups score near 0 percent on a learning outcome test. As another example, suppose the lesson is on adding and subtracting fractions, but the test measures the adding and subtracting of whole numbers. These illustrations show that even when researchers seek to measure learning outcome, they may err by creating an insensitive measure—one that is not appropriate for the level of learners or the to-be-learned material. Some pilot testing may be needed to identify sensitive measures of learning outcome. A somewhat-related problem arises when the assessment test is misaligned with the instructional objective of the game, such as using a general reading achievement test when the game focuses on teaching a much narrower and more specific set of skills.

In taking an evidence-based approach to game research, careful attention must be paid to the quality of the evidence. Designing appropriate dependent measures is a crucial first step in being able to provide high-quality evidence. Developing appropriate dependent measures can be the most challenging aspect of conducting game effectiveness research. My advice is to use multilevel assessments of what was learned that range from retention (i.e., being able to remember the essential information) to transfer (i.e., being able to use the learning information to solve new problems, adapt to new situations, and engage successfully in new learning). Although retention tests are usually easier to develop, transfer tests offer an indispensable indication of how well the learner understands the material (Mayer, 2011a).

Some appropriate techniques for administering learning outcome tests are using the posttest score, pretest-to-posttest gain, and embedded test score. First, sometimes learning outcomes can be measured on a posttest administered immediately after learning, or even better, after a delay of a week or more. Delayed tests are particularly useful in determining the persistence of learning outcome effects.

Second, sometimes learning outcome can be measured as a pretest-to-posttest gain. A serious methodological problem, though, is that taking a pretest can serve as a learning episode—which has been called the *testing effect* (Johnson & Mayer, 2009; Roediger & Karpicke, 2006). Thus, it may be useful to minimize the length of the pretest or use some other preliminary measure, such as a questionnaire.

My advice is also to use embedded assessments whenever possible—that is, assessments that occur as a natural part of the game, perhaps as a special level of the game that requires remembering or applying what has been learned, a request to explain something to a game character as part of the game action, or even a brief request to answer a question or select an answer from a list as an interlude within the game.

Don't Bother to Include a Control Group

People who play game X—a highly engaging first-person shooter game—for twenty hours show an amazing increase in their score on spatial ability tests. People who play game Y—a highly engaging physics simulation game—show a 50 percent improvement in their knowledge of physics from pretest to posttest. After playing game Z—a highly engaging adventure game aimed at teaching scientific reasoning—almost all students report that they are interested in a career in science. These all might appear to be interesting pieces of research evidence, but on closer inspection, each is seriously flawed because no control group was included.

This situation violates the criterion of *experimental control*—the idea that the treatment group should be compared to a control group that is equivalent in all respects except the independent variable. When there is no control group, it is not logically possible to attribute the change in knowledge to the game. What would you say if the same increase in spatial ability testing could be achieved by simply giving the tests multiple times, or if asking students to read a textbook lesson resulted in a 50 percent improvement in physics knowledge, or if viewing a video about the biography of a scientist also resulted in most students wanting to pursue a career in science. As you can see, in game effectiveness research we are not interested in whether the game group showed an improvement but rather we are interested in whether the game group showed a significantly larger improvement than an appropriate control group. Similarly, we are not interested in whether the game group scores well on a postquestionnaire or posttest but instead we are interested in whether the game group scores significantly higher than an appropriate control group. In short, in order to counter the urge to conduct useless game studies, researchers need to remember the motto: "No control group, no gain."

Compare Apples to Oranges

Even when a control group is included, it might not actually offer much experimental control. The third technique listed in table 2.5 is to compare apples to oranges. Consider a study in which the game group plays a math adventure game twenty minutes a day for a month, whereas a control group engages in regular classroom activity in which students learn to solve different math problems than in the game. This study somewhat violates the criterion of *experimental control*—the idea that both groups should be equivalent on all characteristics except the one being tested in the study (e.g., game versus conventional media).

The solution is to equate the groups as much as possible on all features except the inclusion of a game. Both groups should receive identical content, including the same problems and same feedback. Both groups should learn under the same conditions, such as in class every day from 10:00 to 10:20 a.m. Both groups should be tested under the same conditions, such as on the same day by the same teacher. Overall, data are useful to the extent that they come from well-controlled experiments.

Use Preexisting Groups

Consider a study in which one class is selected to be in the game group and another class is selected to be in the control group. When the results come in, the control group shows a greater pretest-to-posttest gain than the game group. This looks bad for the instructional effectiveness of the game, right? On closer inspection, you find that the game group was in an honors class where the average pretest score was high, whereas the control group was in a regular class where the average pretest score was low. Therefore, the control group had much more room to demonstrate a gain than did the game group, thereby making the gain scores nearly impossible to interpret. This study violates the criterion of *random assignment*—the idea that each participant is selected to be placed in the treatment or control group based on chance.

The solution to this problem is to randomly assign students to groups rather than to use preexisting groups. If the groups differ on an important characteristic in spite of random assignment, such as their pretest score or the percentage of boys and girls, this variable can be statistically removed such as by using analysis of covariance.

Consider another example where a group of students that plays action video games at least twenty hours a week scores higher on a test of spatial ability than a group that spends less than one hour per week playing action video games. Although it might be tempting to conclude that game playing is good for cognitive ability, you really cannot draw that conclusion from this study. The problem is that the participants self-selected their groups rather than being randomly assigned, so it is possible that people who have good spatial skills like to spend time playing action video games. In order to make causal conclusions, we need to find studies in which the participants were randomly assigned to groups instead of being self-selected.

Skimp on the Number of Participants

Not only must the participants be randomly assigned to groups, there must be enough of them assigned to each group to matter. Suppose you read a research report in which students who play an enhanced version of a game score 10 percentage points better on a classroom quiz than those who play the base version of the game, effectively increasing their grade by one letter. As you read on, however, you find that there were only eight students in each group, and the difference in their quiz scores was not statistically significant and the effect size was small. This study lacks *statistical power*—the idea that there must be enough participants to adequately test the hypothesis.

This is an example of an underpowered study—one in which too few participants were tested. Although the trend of the data is encouraging, the study is somewhat useless, because we do not know how to interpret the results. According to Cohen (1988), we would need at least twenty-five participants in a group if we expect a large effect and a hundred per group if the expected effect is in the small range. In the world of educational research, bad data (based on underpowered studies) can be more distracting than no data at all. In short, studies with fewer than twenty-five participants in each group should be viewed with extreme caution.

Rely on Anecdotes

Consider the following quote from the parent of a game player, as reported by Prensky (2006) in *Don't Bother Me Mom—I'm Learning*: "I am a mom with a teenage son and I have found that he desperately wanted me to understand why his online games were so appealing to him and for me to understand that he learns from each and every game he plays.... As I watched and asked a multitude of questions, I came to understand the complexity of games and why they are so appealing to him" (p. 146–147). The mom goes on to explain that in discussing strategies with other players, her son is improving his cognitive skills.

What's wrong with fascinating interviews with game players and with people who know them? Such reports can provide richness to our understanding of how game playing affects people, but they are not useful data in assessing game effectiveness. In other words, anecdotes are no substitute for hard data when your goal is to determine the instructional effectiveness of a game. Like the first useless approach in table 2.5, relying on anecdotes violates the criterion of appropriate measures.

Use an Ineffective Game

Suppose a researcher finds that students learn more poorly from playing a math quiz game than from a classroom lesson covering the same material. Nevertheless, when you look more closely at the game, you see that it has an unappealing interface with lots of irrelevant flying objects, annoying sounds, and an incomprehensible game theme. In short, it is simply not a good game. What the study basically has found is that a bad game is not as effective as a good classroom lesson, which is not really a useful piece of information.

This situation involves an inappropriate independent variable, thereby violating the criterion of *experimental control*. The remedy is to make sure that the games being used have been field tested to be appealing for the participants and well designed in terms of providing access to the academic content.

A similar twist on this theme is to implement a good game in an ineffective way, such as asking students to play it in a computer lab during their recess or not relating the game to ongoing activity within the classroom. The remedy is to make sure that games are incorporated well within the school, work, or home context.

Do Not Report on Means, Standard Deviations, and Sample Size

Even if a study uses an appropriate dependent measure, it can become useless by failing to precisely report the scores on that measure. Consider the bar graph shown in figure 2.1 in which the bar for the treatment group is

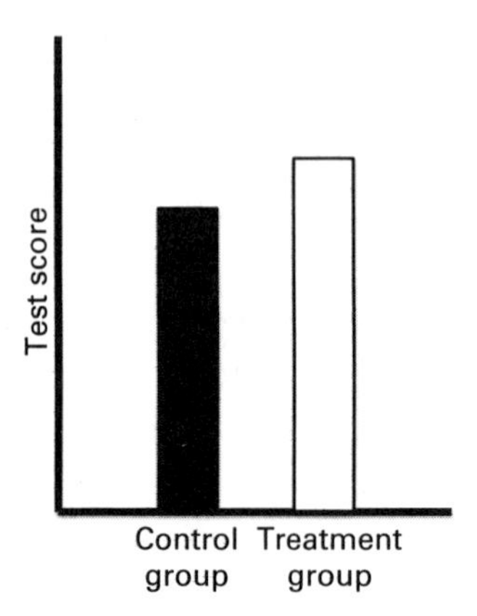

Figure 2.1
A graph that does not provide enough information

higher than the one for the control group. What is wrong with this artistic rendering of the data? If this is the only presentation of the data, the problem is that it fails to provide us with a precise number for the mean (*M*), standard deviation (*SD*), and sample size (*n*) of each group, which is the minimal information needed to compute the effectiveness of the treatment.

Alternatively, consider a research report that supplies no data graphs or tables, and instead simply tells us the bottom line: "A t-test revealed that the posttest score of the game group was significantly greater that the post-test score of the control group, at $p < 0.05$." What is wrong with this seemingly concise statement? Again, the problem is that it fails to provide us with the mean, standard deviation, and sample size of each group. For scholars interested in what the research evidence has to say about game effectiveness, there is nothing more discouraging than reading through a well-done study on game effectiveness only to find that the essential means and standard deviations are missing.

The reason these numbers are important is that they are needed to compute a value of *effect size*—how many standard deviations the experimental group differs in comparison to the control group. A study can meet all the criteria and look like a perfect example of game research, but if there is no way to compute effect size, it becomes less than useful as a source of evidence concerning game effectiveness. The central role of effect size is described in the next section.

Role of Effect Size in Game Research

An important metric for examining the instructional effects of a game is effect size (Cohen, 1988; Ellis, 2010; Lipsey & Wilson, 2001; Rosenthal, Rosnow, & Rubin, 2000). In game research, effect size (*d*) can be computed by subtracting the mean learning outcome score of the control group from the mean learning outcome score of the treatment group, and then dividing the result by the pooled standard deviation:

$$\text{effect size} = \frac{(\text{mean score of treatment group} - \text{mean score of control group})}{(\text{pooled standard deviation of both groups})}$$

For example, on a posttest measuring the academic content of a game, if $M = 4.7$ and $SD = 1.0$ for the treatment group, and $M = 3.9$ and $SD = 1.0$ for the control group, then the effect size is $d = (4.7 - 3.9)/1.0 = 0.8/1.0 = 0.8$. Cohen (1988) has suggested that effect sizes below $d = 0.2$ are negligible, at $d = 0.2$ are small, at $d = 0.5$ are medium, and above $d = 0.8$ are large. Hattie (2009) has proposed that any effect size greater than $d = 0.4$ is educationally important. In game research, a good rule of thumb is to focus on game effects that are above $d = 0.4$, and to be particularly impressed with game effects that are greater than $d = 0.8$. In some cases, though, even a small effect size may be practically important. In addition to the practical significance indicated by a large effect size, the effect should be statistically significant at the $p < 0.05$ level based on traditional measures of significance such as t-tests or analyses of variance.

On the positive side, effect size offers a common metric for comparing the strength of an instructional treatment across different studies. Effect size is a simple measure that is easy to communicate and understand, because it tells us how many standard deviations better (or worse) one group scored as compared to another. On the negative side, reducing an entire research report to one number requires losing a substantial amount of information that might be relevant to understanding game effectiveness.

On balance, I have opted to concentrate on effect size because of its usefulness in conducting the meta-analyses reported in chapters 4 through 7. In a meta-analysis, the effect sizes from all experimental comparisons that share the same experimental design (such as value-added, cognitive consequences, or media comparison design) are summarized as an average effect size. In the meta-analyses reported in this book, I focus on the *median effect size*—the effect size that is greater than half the studies and less than half the studies.

Note

Portions of this chapter were adapted from Mayer & Lieberman (2011).

References

Cohen, J. (1988). *Statistical analysis for the behavioral sciences* (2nd ed.). Hillsdale, NJ: Erlbaum.

Ellis, P. D. (2010). *The essential guide to effect sizes*. New York: Cambridge University Press.

Hattie, J. (2009). *Visible learning: A synthesis of over eight hundred meta-analyses relating to achievement*. New York: Routledge.

Johnson, C. I., & Mayer, R. E. (2009). A testing effect with multimedia learning. *Journal of Educational Psychology, 101*, 621–629.

Lipsey, M. K., & Wilson, D. B. (2001). *Practical meta-analysis*. Thousand Oaks, CA: Sage.

Mayer, R. E. (2009). *Multimedia learning* (2nd ed.). New York: Cambridge University Press.

Mayer, R. E. (2011a). *Applying the science of learning*. Upper Saddle River, NJ: Pearson.

Mayer, R. E. (2011b). Multimedia learning and games. In S. Tobias & J. D. Fletcher (Eds.), *Computer games and instruction* (pp. 281–306). Charlotte, NC: Information Age Press.

Mayer, R. E., & Lieberman, D. (2011). Conducting scientific research on learning and health behavior change with computer-based health games. *Educational Technology, 51*(5), 3–14.

Mosteller, F., & Boruch, R. (2002). *Evidence matters: Randomized trials in educational research*. Washington, DC: Brookings Institution.

O'Neil, H. F., Wainess, R., & Baker, E. (2005). Classification of learning outcomes: Evidence from the game literature. *Curriculum Journal, 16*, 455–474.

Phye, G. D., Robinson, D. H., & Levin, J. (Eds.). (2005). *Empirical methods for evaluating educational interventions*. San Diego, CA: Academic Press.

Prensky, M. (2006). *Don't bother me mom—I'm learning.* Saint Paul, MN: Paragon House.

Roediger, H. L., & Karpicke, J. D. (2006). The power of testing memory: Basic research and implications for educational practice. *Perspectives on Psychological Science, 1*(3), 181–210.

Rosenthal, R., Rosnow, R. L., & Rubin, D. B. (2000). *Contrasts and effect sizes in behavioral research.* New York: Cambridge University Press.

Shavelson, R. J., & Towne, L. (Eds.). (2002). *Scientific research in education.* Washington, DC: National Academy Press.

3 Theory: Applying Cognitive Science to Games for Learning

Chapter Outline

Summary

If we want to create or select effective games for learning, it is useful to understand how learning works, how assessment works, and how instruction works. Learning takes place within the human information-processing system, which has separate channels for visual and verbal material, three memory stores including a working memory with limited processing capacity in each channel, and cognitive processes such as selecting, organizing, and integrating information. Assessment involves specifying what we want the player to learn (i.e., the learning objective) and developing ways to measure what was learned (i.e., the learning outcome) that are valid, reliable, objective, and referenced. Instruction may be aimed at reducing extraneous processing (which is cognitive processing that does not serve the learning objective), managing essential processing (which is cognitive processing aimed at mentally representing the essential material), and fostering generative processing (which is cognitive processing aimed at making sense of the essential material). Theories of learning are relevant to acquiring facts (e.g., reinforcement theory), constructing schemas (e.g., schema theory), automatizing procedures (e.g., automaticity theory), and building strategies (e.g., social learning theory). Theories of academic motivation focus on the learner's interests (e.g., interest theory and value theory), beliefs (e.g., self-efficacy theory, attribution theory, and self-theories), goals (e.g., goal orientation theory), and needs (e.g., self-determination theory and intrinsic motivation theory). The challenge of effective game design is to use game features that promote motivation to learn, but do not disrupt the appropriate cognitive processing during learning; and to use instructional features that prime appropriate cognitive processes during learning, but do not shut down the player's motivation to learn.

Applying Cognitive Science to Learning with Games

Science of Learning: How People Learn

If we are interested in how people learn with games, it is worthwhile to consider how learning works. The science of learning is the scientific study of how people learn (Mayer, 2011). In the context of games for learning, the science of learning is the scientific study of how people learn with games. In this section, I describe aspects of the science of learning that are most relevant to educational games.

Let's begin with a description of the human information-processing system, which is the system humans use for learning. Figure 3.1 summarizes the human learning system in a way that is relevant to game learning, as depicted in the cognitive theory of multimedia learning (Mayer, 2009, 2011). The model in figure 3.1 reflects three research-based principles about the human information-processing system derived from cognitive science:

Dual channels People have separate channels for processing auditory/verbal versus visual/pictorial material (Baddeley, 1992, 1999; Paivio, 1986, 2006). As you can see, the top row in the figure depicts the auditory/verbal channel, beginning with information entering through the ears, and the bottom row depicts the visual/pictorial channel, beginning with information entering through the eyes.

Limited capacity People can engage in only a small amount of processing in each channel at any one time (Baddeley, 1992, 1999; Sweller, 1999; Sweller, Ayres, & Kalyuga, 2011). The box labeled "Working memory" in

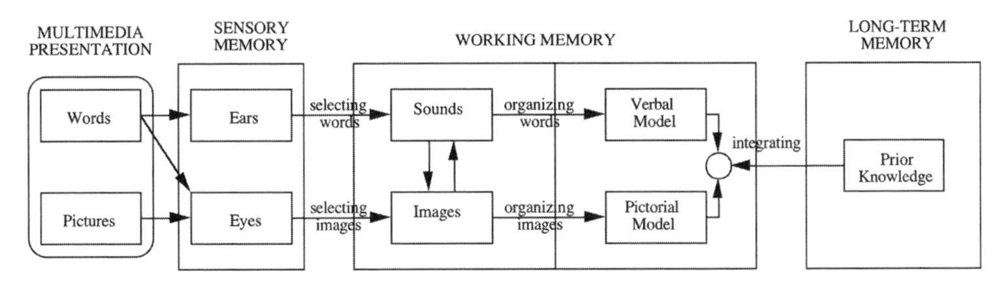

Figure 3.1
A cognitive theory of multimedia learning

figure 3.1 is the bottleneck in the system in which only a few items can be processed at any one time within each channel.

Active processing Meaningful learning occurs when people engage in appropriate cognitive processing during learning, including attending to relevant material, mentally organizing it into a coherent representation, and mentally integrating it with relevant prior knowledge (Mayer, 2009; Wittrock, 1989). The arrows in figure 3.1 represent the cognitive processes of selecting, organizing, and integrating.

This analysis shows that to understand how games can promote learning, we begin by recognizing that learners need to engage in a lot of appropriate processing (i.e., active processing) about words and pictures during game play (i.e., dual channels), but they are extremely limited in the capacity they have to do so (i.e., limited capacity).

The boxes in figure 3.1 represent three types of memory stores:

Sensory memory Unlimited sensory input from the ears or eyes is held in sensory form for a very brief period of time (e.g., a quarter of a second).

Working memory Information transferred from sensory memory is converted into verbal and pictorial representations that can be mentally manipulated, but only a few pieces of verbal material and a few pieces of pictorial material can be processed at any one time. We can be consciously aware of cognitive processing in working memory, but the material decays within about twenty seconds if we do not actively process it.

Long-term memory The learner's permanent storehouse of knowledge and skill is located in long-term memory, which through meaningful organization has unlimited capacity.

As you can see, both sensory memory and long-term memory have unlimited capacity, but working memory is severely limited. This structural feature of the human information system has crucial implications for learning with games, because it is easy to overload the game player's working memory, thereby decreasing the opportunity for making sense of the material.

The arrows in figure 3.1 represent three types of cognitive processes during learning:

Selecting Auditory information from the game results in an auditory sensory representation being held in auditory sensory memory, and visual information from the game results in a visual sensory representation being held in visual sensory memory. If the player does not actively attend to those representations, they will quickly disappear from the game player's cognitive system. If the player pays attention to pictorial material in the game (indicated by the "selecting images" arrow), it will be transferred to working memory for further processing. If the player pays attention to auditory material in the game (indicated by the "selecting sounds" arrow), it will be transferred to working memory for further processing.

Organizing Next, the player can mentally organize the incoming sounds into a coherent verbal representation (indicated by the "organizing sounds" arrow) and can mentally organize the incoming images into a coherent pictorial representation (indicated by the "organizing images" arrow).

Integrating Finally, the player can mentally combine the verbal and pictorial representations with each other as well as with relevant prior knowledge activated from long-term memory (indicated by the "integrating" arrows).

Once new knowledge is constructed, it can be transferred to long-term memory for permanent storage. This storage process could be indicated by an arrow from working memory to long-term memory.

Table 3.1 lists three cognitive processes that are needed for meaningful learning with games—selecting, organizing, and integrating. For example, consider playing an arithmetic simulation game in which a bunny hops along a number line to correspond to number expressions such as 3 – (–5) = _____ (Moreno & Mayer, 1999). The process of selecting is reflected in noticing that the bunny starts at the 3 position, the bunny says "3," the

Table 3.1
Three cognitive processes for meaningful learning

Process	Description
Selecting	Paying attention to relevant incoming information
Organizing	Mentally arranging the selected information into coherent cognitive representations
Integrating	Mentally connecting the cognitive representations with each other and with relevant prior knowledge activated from long-term memory

bunny turns 180 degrees backward to face the left side of the number line, the bunny says "minus," the bunny jumps backward 5 steps, and the bunny says "negative 5." The process of organizing is reflected in mentally building the auditory sequence "3" "minus" "negative 5," and pictorial sequence on the number line of starting at 3, turning around, and jumping backward 5 steps. The process of integrating is reflected in remembering how a number line works, and connecting "3" with the bunny starting at the 3 position, "minus" with turning around, and "negative 5" with jumping backward 5 steps.

What are the implications of the cognitive theory of multimedia for learning with games? We want to know how game features and instructional features affect these three kinds of cognitive processing during learning. Overall, meaningful learning with games requires that the learner engage in selecting, organizing, and integrating, but game mechanics can easily distract the player from these cognitive processes. If the goal of game playing is to help learners build learning outcomes based on instructional objectives, designers may need to add instructional features that guide learners' cognitive processing during game play. The effectiveness of adding instructional features is examined in chapter 5 on the value-added approach to game research.

Science of Assessment: How Assessment Works

The study of how people learn with games depends on having useful ways to measure learning outcomes. The science of assessment is the scientific study of how to determine what people know (Mayer, 2011; Pellegrino, Chudowsky, & Glaser, 2001). As shown in table 3.2, the goal of the science of assessment is the creation of instruments for assessing learning outcomes

Table 3.2
Four characteristics of a useful assessment

Characteristic	Description
Valid	Assessment serves the intended purpose
Reliable	Assessment gives the same score every time
Objective	Assessment is scored the same way by all scorers
Referenced	Assessment yields score that is interpretable

Source: Adapted from Mayer (2011).

that are *valid, reliable, objective,* and *referenced.* Consider a game that teaches students how electric circuits work on levels one through nine, and then provides a twenty-five-item embedded transfer test on level ten consisting of new electric circuit problems using the same eight principles of electricity as taught in the game (Johnson & Mayer, 2010; Mayer & Johnson, 2010).

As shown in the first line of table 3.2, the assessment is valid to the extent that the goal is to assess the game player's knowledge of the principles of electricity flow. If our assessment focused on the structure of atoms or the history of Ohm's law, it would not be valid because the game was intended only to teach the principles of electron flow in circuits. If our assessment consisted of asking game players how much they liked the game or even how much they think they learned from it, we again have questionable validity because self-reports are generally not as useful as measures that require actual performance (such as recalling the material or solving a problem based on the material). One way to gauge the validity of a game assessment is to compute the correlation between the score on the game assessment with a criterion such as a school grade or achievement test score on the same topic. If the correlation is positive and high, we have some evidence of validity; if not, we may be working with a measure that lacks validity.

As shown in the second line of table 3.2, the assessment is reliable to the degree that it consistently gives the same score. One way to determine the reliability of our assessment is to see how half the test correlates with the other half, using a statistic called Cronbach's alpha. If alpha is large, we have evidence that the test items all measure the same thing; if not, we can't be sure about what is being assessed.

As shown in the third line of table 3.2, our level-ten assessment is objective because each item is easily scored as right or wrong, and there is no room for subjective judgment in counting up the number of correct answers. If we used an open-ended assessment, such as asking players to write a description of how circuits work, we would need a precise scoring rubric—that is, a scoring key that specifies exactly how to score the answers. One way to determine objectivity is to compute the correlation between the scores generated by two independent raters. If the interrater correlation is positive and high, we have evidence of objectivity; if not, there is reason to question whether the assessment is objective.

Finally, based on the last line of table 3.2, our assessment is referenced because we can tell where each player (or group) stands with respect to the number of standard deviations above or below the mean, yielding what is called a z-score. This is an example of *norm referencing* because each score can be compared to the other test takers. In *criterion referencing,* test performance is compared to a specific learning objective, such as obtaining 80 percent correct on target tasks. Overall, the entire enterprise of studying games for learning relies on learning outcome measures that are valid, reliable, objective, and referenced. As described in chapter 2, one of the major characteristics of useful research on games for learning is the inclusion of appropriate assessments of learning outcome.

Table 3.3 lists three functions of assessments: *preassessment,* which is conducted before game playing to determine what the player already knows, and which can be used to adjust the game level; *formative assessment,* which is conducted during game playing to determine how well the player is learning, and can be used to adjust the amount of support provided during game play; and *summative assessment,* which is conducted after game playing to determine what was learned, and can be used in making decisions about redesigning the game or the way it is used. Most of the studies examined in this book focus on summative assessments (i.e., posttest scores), or the difference between preassessment and summative assessments (i.e., pretest-to-posttest gain scores). In the future, intelligent games for learning may include more opportunities for adjusting the initial game level based on preassessment and continually adjusting game conditions based on formative assessments that are embedded in the game.

What is a learning outcome? A *learning outcome* is a change in knowledge caused by instruction. In terms of games for learning, a learning outcome is

Table 3.3
Three functions of assessments

Name	When	Description
Preassessment	Before game playing	To determine what the game player already knows (or the player's characteristics)
Formative assessment	During game playing	To determine what the game player is learning during playing the game
Summative assessment	After game playing	To determine what was learned from playing the game

Table 3.4
Four kinds of academic knowledge that can be learned in games

Type of knowledge	Elements	Description	Example	Typical game
Factual knowledge	Facts	Characteristics or states of components	Ohm discovered Ohm's law	Quiz
Conceptual knowledge	Schemas, models, categories, or principles	Generalized cognitive structures	I = V/R	Simulation
Procedural knowledge	Procedures	A step-by-step process	50/2 = 25	Action
Strategic knowledge	Strategies	A general method	Find a related problem	Role-playing

the knowledge that the learner has constructed during the learning episode, such as during playing a game like the *Circuit Game* described above. Table 3.4 lists four kinds of knowledge identified by cognitive and educational scientists that are most relevant to game playing: *factual knowledge*, such as knowing that Georg Ohm discovered Ohm's law or that Ohm's law is about electric flow in a circuit; *conceptual knowledge*, such as knowing the principle that the intensity of electron flow is directly proportional to voltage and inversely proportional to resistance; *procedural knowledge*, such as knowing the arithmetic procedures for multiplication and division; and *strategic knowledge*, such as knowing that if you can't solve the presented problem, you should think of a related problem you can solve (Anderson et al., 2001; Mayer, 2011). As you can see in the last column of table 3.4, quiz games may be used to teach facts; simulation games may be used to help learners build concepts; action games may be used to build procedures; and role-playing adventure games may be used for building strategic knowledge. Assessments of game learning should focus on one or more of these kinds of knowledge.

What is not an assessment of learning outcome? Assessments of how much a player likes a game are not appropriate measures of learning outcome because liking does necessarily equate to learning. Assessments based on describing the game player's keystrokes during game play are not appropriate measures of learning outcome because activity does not necessarily equate to learning. Assessments based on asking game players to rate how much they learned are poor measures of learning outcome because people often lack awareness of their learning. Although each of these measures may have some value in addressing specific research questions, they are not substitutes for assessments of learning outcome.

Assessments of learning outcome should be keyed to specific learning objectives. A *learning objective* is a specification of the intended change in the learner's knowledge. In the case of games for learning, the change is caused by game playing. When an assessment item matches a learning objective, we can say it is *aligned*, such as when we assess the *Circuit Game* by asking players to solve similar problems to those given in the game. When an assessment item does not correspond to any of the learning objectives, we can say it is *not aligned*, such as when we test players on material that was not covered in the game. Having nonaligned test items violates the criterion of validity listed in table 3.2, and therefore nonaligned test items are not useful in research on games for learning.

It is useful to distinguish between two ways to measure learning outcomes caused by game playing: *retention tests*, aimed at assessing what the game player remembers; and *transfer tests*, aimed at assessing what the game player understands. As shown in table 3.5, after playing the *Bunny Game* discussed earlier in this chapter, we can ask the game player to solve the same arithmetic problems as in the game (i.e., retention test) or different arithmetic problems based on the same underlying principles (i.e., transfer test). I am most interested in whether games can be a vehicle for

Table 3.5
Two ways to measure learning outcomes

Type of test	Target of test	Description	Example
Retention	Remember	Recall or recognize what was presented, or solve same problem as was presented	Knowing that "–5" is read as "negative 5"; after seeing how to solve 3 – (–5) = in the *Bunny Game*, solve it again on the test
Transfer	Understand	Evaluate or use presented material in a new situation, or to solve new problems	Being able to explain the difference between a minus sign and negative sign; after seeing how to solve 3 – (–5) = ______ in the *Bunny Game*, solve –3 – 15 = ______

improving understanding, so I prefer to concentrate on measures of transfer wherever possible in reviewing game research in this book (i.e., in chapters 4 through 7).

Learning outcomes can be assessed on *immediate tests*, given immediately after playing the game, or *delayed tests*, given after a retention interval (which can vary from a few minutes to a week to a year or beyond). Although most of the research studies reported in this book involve immediate tests, it is also useful to determine whether the effects persist over time.

Learning outcomes can be assessed in *embedded tests*, which occur during game play and appear to the player as part of the game, or *separated tests*, which take place before or after game play as a separate activity. Although most of the research studies reported in this book involve separated tests, it is also useful to explore ways to embed assessments within game play.

Finally, three common formats for assessment are *forced choice tests*, such as multiple-choice tests or true-false tests; *open-ended tests*, such as recall tests or essay tests; and *performance tests*, such as solving a problem or completing a task. It is possible to test for transfer using any of these formats, although both open-ended and performance tests are generally preferred. In order to meet the criterion of objectivity (as outlined in table 3.2), however, open-ended tests and performance tests need clear and precise scoring rubrics—that is, specifications for how to score them.

An important overall theme of the science of assessment is that studies of game effectiveness need clear specifications of the learning objectives so we can identify what it is that game players are supposed to learn from the game. Based on their learning objectives, game effectiveness studies depend on solid measures of learning outcome that meet the criteria of validity, reliability, objectivity, and referencing.

Science of Instruction: How Instruction Works

Game playing is a somewhat-unique instructional environment, but the science of instruction still provides a useful theoretical framework. The *science of instruction* is the scientific study of how to help people learn (Mayer, 2011). In the context of games for learning,the science of instruction seeks to determine how to help people learn with games.

Based on the model of learning presented in figure 3.1, we begin with the recognition that game players have limited capacity for processing

Table 3.6
Three kinds of cognitive load during game playing

Type of load	Description	Cause
Extraneous	Cognitive processing that does not support the learning objectives	Distracting game features
Essential	Cognitive processing aimed at mentally representing the content	Complexity of the content
Generative	Cognitive processing aimed at making sense of the content	Motivating game features

verbal and pictorial material in working memory, but meaningful learning requires that they select, organize, and integrate relevant verbal and pictorial material. Table 3.6 lists three demands on the game player's limited processing capacity (Mayer, 2009, 2011): *extraneous processing,* which is not related to the learning objective; *essential processing,* aimed at mentally representing the essential academic material (i.e., selecting relevant information and organizing it as presented); and *generative processing,* which is deeper cognitive processing aimed at making sense of the essential academic material (i.e., reorganizing the information and integrating it with relevant prior knowledge). As you can see in the third column of table 3.6, extraneous processing is caused by distracting game design; essential processing is caused by the complexity of the to-be-learned material; and generative processing is caused by the game player's motivation to learn as stimulated by exciting game design. Overall, the challenge of designing games for learning is to reduce distracting features (and thereby reduce extraneous processing), hone motivating features (and thereby foster generative processing), and present academic content in ways that manage essential processing.

Figure 3.2 presents three game learning scenarios (Mayer, 2011): extraneous overload, essential overload, and generative underutilization. Suppose you play a game with attention-grabbing graphics, music, and fast-paced action that is intended to teach you about Newton's laws of motion. Processing all the incoming sensory information and all the outgoing actions may overload your cognitive system (i.e., extraneous processing), leaving little opportunity to figure out what is important or reflect on what is happening (i.e., essential and generative processing). This is an example of *extraneous overload,* in which the demands for extraneous processing are

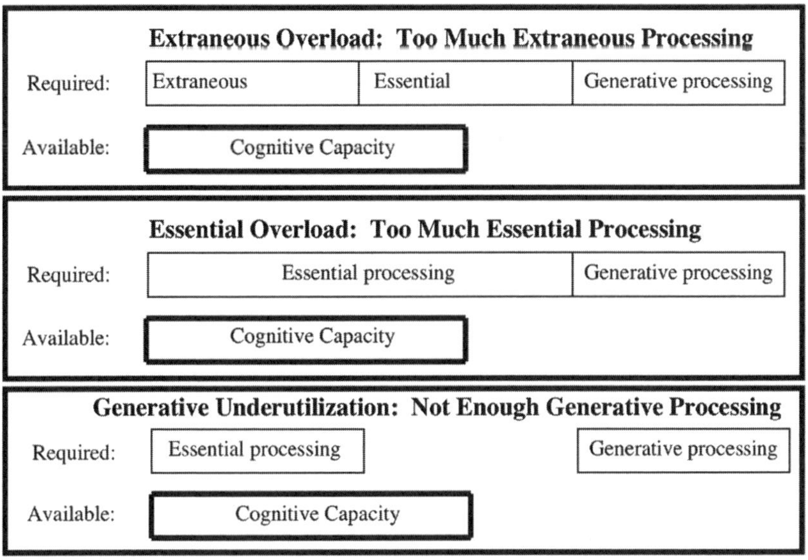

Extraneous Overload: Too Much Extraneous Processing

Required:	Extraneous	Essential	Generative processing
Available:	Cognitive Capacity		

Essential Overload: Too Much Essential Processing

Required:	Essential processing	Generative processing
Available:	Cognitive Capacity	

Generative Underutilization: Not Enough Generative Processing

Required:	Essential processing	Generative processing
Available:	Cognitive Capacity	

Figure 3.2
Three instructional scenarios

so great that there are not enough cognitive resources left for carrying out the needed essential and generative processing. The solution is to reduce extraneous processing by cutting down on the distracting game features.

Next, consider a situation in which we have toned down the distracting features of the game, but the academic content is so complicated that the game player is not able to make sense of it. This is an example of *essential overload,* in which the demands of essential processing are so great that there are not enough cognitive resources left for carrying out all the needed essential and generative processing. The solution is to manage essential processing by explicitly guiding the game player's cognitive processing during game playing.

Finally, consider a situation in which we have toned down the distracting game features and added instructional features for managing essential processing, but game players are not motivated to learn. In our attempts to save the game as an instructional venue, we may have destroyed its motivating properties. In short, the game no longer is much fun. This is an example of *generative underutilization,* in which the game player has the cognitive resources available for deeper cognitive processing, but is not

Table 3.7
Three instructional goals

Instructional goal	Possible implementation
Reduce extraneous processing	Immersion: Reduce perceptual realism Redundancy: Do not add printed words that correspond to spoken words
Manage essential processing	Pretraining: Provide pretraining in key concepts Coaching: Offer advice or explanations Segmenting: Break game into segments Modality: Present words in spoken form
Foster generative processing	Self-explanation: Provide prompts for self-explanation Personalization: Use conversational style Choice: Allow choice of interface appearance

motivated to use them. The solution is to foster generative processing by incorporating entertaining features that do not cause too much extraneous processing.

Table 3.7 lists some possible techniques for accomplishing each instructional goal. In order to reduce extraneous processing, some possible techniques are to reduce perceptual realism (if it is not needed for the learning objective) and eliminate on-screen text that matches spoken text (if the material is short and familiar to the learner). In order to manage essential processing, some possible techniques are to provide pretraining in the key concepts used in the game, include expert advice and explanations throughout the game, break the game into manageable segments, and present words in spoken form as a way to free up capacity for visual processing. In order to foster generative processing, we could add prompts to a game that require the game player to reflect, use conversational style to build a social partnership with the game player, and allow the game player to choose some aspects of the way the interface looks. These theory-based suggestions are put to the test in chapter 5, which examines the value-added approach to game research.

How Games Affect Learning

In order to investigate how games affect learning, an important step is to clearly describe what we want the player to learn—that is, we need to

Table 3.8
Learning theories for four kinds of academic knowledge that can be learned in games

Type of knowledge	Theory	Learning involves	Teaching involves
Factual knowledge	Reinforcement	Strengthening and weakening of associations	Dispensing rewards and punishments for player actions
Conceptual knowledge	Schema	Constructing organized mental representations	Giving guidance in how a system works
Procedural knowledge	Automaticity	Automating a cognitive skill	Giving repeated practice with appropriate feedback
Strategic knowledge	Social learning	Acquiring usable methods for attacking complex tasks	Modeling how to attack complex tasks

describe the to-be-learned knowledge. As summarized earlier in table 3.4, four kinds of knowledge that can be learned in games are facts, concepts, procedures, and strategies. Table 3.8 summarizes the research-based learning theories that are most relevant for learning each kind of knowledge: reinforcement theory for learning factual knowledge; schema theory for learning conceptual knowledge; automaticity theory for learning procedural knowledge; and social learning theory for learning strategic knowledge. However, all four kinds of knowledge are required to achieve proficiency in most complex tasks ranging from solving word problems to comprehending a paragraph. In short, proficient students know lots of relevant facts that are indexed for speedy retrieval when needed, have constructed schemas for organizing their knowledge, have automated their basic cognitive procedures, and know useful strategies for how to approach new tasks. Let's examine each of these in turn, with an eye toward the underlying theories of learning and how they relate to games for learning.

Building Factual Knowledge: Reinforcement Theory

I have to admit that teaching factoids—individual pieces of information—is not my favorite learning goal, but on the other hand, having lots of factual knowledge (that eventually gets integrated) can be an important part of developing proficiency in many academic disciplines. For example, we might want young children to learn the names of the days of the week and we might want older students to learn the atomic number for each chemical element in the periodic table.

When the goal is to help learners build associations (such as between the shape of California on a map and the spoken word "California"), reinforcement theory appears to be an important theoretical framework. The main idea in reinforcement theory, immortalized as E. L. Thorndike's (1911) *law of effect*, is that behaviors that are followed by satisfaction to the learner are more likely to be repeated in the future under the same circumstances, and behaviors that are followed by dissatisfaction to the learner are less likely to be repeated in the future under the same circumstances. Rewards cause a strengthening of the association between the situation and successful behavior; punishments cause a weakening of the association between the situation and unsuccessful behavior.

For example, in playing an electronic quiz game suppose you have to click on the country on a map of the world that is on the screen when an on-screen character from the game says, "Where in the world is Australia?" If you click on the correct country, you hear applause and move on to Australia, which gets you closer to winning the game. If you click on the wrong country, you hear a buzzer as the correct answer turns red on the screen with "Australia" printed on it, and you get moved one step back. In this case, according to reinforcement theory, the association between "Australia" and the location of Australia on the map is strengthened if you give the right answer, and the association between "Australia" and whatever incorrect country you selected is weakened if you give the wrong answer. In this version of the game, we have also provided external feedback by showing players where Australia is if they get it wrong.

Reinforcement theory was the first major theory of learning in psychology and education, introduced by the world's first educational psychologist, Thorndike (1911), in the early 1900s, and later popularized in a somewhat extreme form by B. F. Skinner (1938, 1968). Does it still have relevance for today's games for learning? On the positive side, more than a hundred years of research (initially mainly with laboratory animals) have demonstrated the undeniable role of rewards and punishments in learning, and the law of effect is one of the pillars of the psychology of learning. On the negative side, when reinforcement theory is applied to human learning of academic material, it is necessary to broaden it to consider the learner's cognitive interpretation of the learning situation, and in particular, the learner's motivation to learn. In addition, in many academic domains, developing expertise involves integrating facts with each other to form schemas and with other kinds of knowledge to enable successful performance on challenging tasks, so too much of a focus on isolated facts eventually can become counterproductive. In short, the view of learning as response strengthening has a place in the learning sciences, but its role is limited because there is more to academic learning than only forming associations (Mayer, 2009, 2011).

Building Conceptual Knowledge: Schema Theory

In some learning situations, the goal is to help learners construct new schemas—organized knowledge structures used for assimilating new experiences and guiding future behavior. In its broadest sense, schema theory includes all forms of integrated conceptual knowledge, such as concepts, categories, schemas, models, and principles. Suppose, for example, we want students to understand how a cause-and-effect system works, such as how the human circulatory system works (including the role of the heart, lungs, veins, arteries, etc.). According to schema theory, learning involves building a mental model of the parts and the relations among them, so you can make predictions about what would happen under different circumstances.

Alternatively, suppose we want students to learn how electric circuits work (including batteries in series or parallel, resistors in series or parallel, wires, electron flow, etc.) based on Ohm's law. According to schema theory, learning involves building a mental model of the parts in a circuit and the relations among them in way that is consistent with the underlying principles in Ohm's law.

Finally, consider a high school student who can read an arithmetic word problem, and promptly determine whether it is a work problem, time-rate-distance problem, mixture problem, and so on. As you can see, learning the categories of word problems is another form of building schemas. According to schema theory, developing expertise in a field includes learning categories and concepts.

For example, in playing a simulation game, suppose you had to construct various electric circuits by dragging and dropping batteries and resistors to increase or decrease the rate of electron flow within the context of accomplishing some goal in the game. You can see what happens to the rate of flow of electrons when you change two batteries from series to parallel, or when you change from having one resistor to having two resistors in a series. In this way, you develop an understanding of the basic principles underlying how changes in the parts of a circuit affect the rate of electron flow, and thereby you can construct a mental model of the cause-and-effect system.

Building schemas is considered the hallmark of developing expertise in a field (Mayer, 2009; Sweller et al., 2011). When the goal of instruction is to help learners construct conceptual knowledge, schema theory is particularly relevant. Early proponents of schema theory include Jean Piaget (1926), who found that children develop progressively more complex schemas as they gain experience, and Frederic Bartlett (1932), who found that people's prior knowledge affects how they learn and remember verbal material. In short, learning is a process of assimilating incoming information into existing schemas and accommodating incoming information by creating new schemas when needed. In many ways, schema theory (or the broader theoretical framework of constructivism) is the second major conception of how learning works to find popularity in psychology and education, mainly since the 1960s (Mayer, 2009, 2011).

Does schema theory have a place in games for learning? Given the importance of schemas in the learner's development of expertise in a discipline, schema theory remains an important theoretical framework for guiding game development. However, on the negative side, an essential caution is that pure discovery methods—that is, hands-on activity without guidance—generally are inefficient and ineffective ways to promote conceptual learning (Kirschner, Sweller, & Clark, 2006; Mayer, 2004), so simply allowing players to interact with a simulation may not be as effective as providing guidance and instruction.

Building Procedural Knowledge: Automaticity Theory

Suppose the learning objective is to help someone develop a cognitive skill, such as how to read words (such as *dog*) or compute answers for two-column subtraction problems (such as 35 - 17 = ____). These are examples of procedural knowledge, as listed on the third line of table 3.8.

Research on skill learning has a long history in psychology and education, also dating back to Thorndike's (1911, 1931) early work on the effectiveness of practice with feedback. Today, feedback is recognized as an extremely powerful instructional technique in the field of education (Hattie, 2009). In addition to building schemas (as described in the previous section), becoming an expert in a field requires the development of automated procedures—that is, how-to-do-it processes that do not require conscious attention when you use them. For example, when you read, you do not want to have to consciously think about how you decipher each word, and when you are solving a complex math problem, you do not want to waste your attention on how to add two numbers. When you type an essay, you do not want to consciously think about where your fingers are on the keys.

According to theories of skill learning, formalized in the 1950s and beyond, people develop skills in three stages (Fitts & Posner, 1967; Singley & Anderson, 1989):

Cognitive stage A procedure is encoded mainly in declarative form, such as "s makes the sound of a punctured tire" in word reading or "don't forget to carry your tens" in addition.

Associative stage A procedure is encoded as a step-by-step process, but you need to think about it as you apply it.

Autonomous stage A procedure is encoded as a step-by-step process, and you do not have to think about it as you apply it, such as reading a familiar paragraph.

When the goal of instruction is to help students develop automated procedures, automaticity theory comes into play. Research on developing automated procedures shows that learners need to engage in repeated practice with feedback. Ericsson, Charness, Feltovich, and Hoffman (2006) have shown that the development of expertise is best served by *deliberate practice*—in which learners work on tasks that are slightly beyond their current level of competence.

Does automaticity theory have a place in games for learning? Action games, with progressively more challenging levels, appear to offer a potentially powerful venue for developing automated cognitive skills. For example, in a first-person shooter game, as players navigate through game screens, they are constantly confronted with suddenly appearing or fast-moving objects, some of which are threatening and need to be shot. As examined in chapter 6 on the cognitive consequences approach to game research, playing first-person shooter games may be effective in promoting perceptual attention skills that are needed in the sciences and engineering. An important limitation concerns the degree of transfer afforded by playing an action game, because we run the risk that players may automate a very specific set of cognitive skills.

Building Strategic Knowledge: Social Learning Theory

In addition to acquiring lots of readily accessible facts, constructing schemas for organizing information, and automating their basic cognitive procedures, learners need an arsenal of strategies for using their knowledge when attacking new problems. Procedures are step-by-step processes that can be learned through practice with feedback, but strategies are general approaches that need to be carefully selected and adapted to new situations. Learners need strategies for how to orchestrate their knowledge in service of solving a new problem or accomplishing a new task, and they need metastrategies for how to choose, monitor, and revise a strategy.

How do people learn such strategies? According to social learning theory, a well-established theory of learning popularized by Bandura (1986), people learn what to do by watching what more experienced people do. An important instructional implication is that an effective instructional method for teaching cognitive strategies is *modeling*—in particular, comparing how a successful person addresses a task versus how the learner does it.

In the context of games for learning, games that allow for role-playing within challenging scenarios or cases appear to offer a way of helping people develop strategic knowledge. In a role-playing game, for example, a player can take the role of a factory manager who must make decisions about how to set up an assembly line to maximize efficiency. An on-screen agent offers advice, when needed, and even models how to attack parts of the problem. A potential danger with role-playing games is that dealing with the game mechanics may require so much time and energy that more direct methods of instruction are more effective. Chapter 7 examines media comparison studies in which learning by playing a game is compared to learning with conventional media such as a slideshow presentation.

Overall, this section explored four visions of learning: learning as the strengthening and weakening of associations; learning as constructing cognitive structures; learning as automatizing procedures; and learning as the modeling of solution methods. Each vision is research based and may be most relevant for a specific kind of learning outcome. Decisions about the design and use of games for learning can be guided by one or more of these learning frameworks, depending on the kind of to-be-learned knowledge.

How Games Affect the Motivation to Learn

Games are widely recognized for their motivational appeal, as evidenced by the popularity of commercial games (Kapp, 2012). In this section, I examine the role of motivation in games for learning, with an eye toward how to harness the motivational power of games for educational purposes. In particular, I examine what the major research-based motivational theories have to say that may be relevant to the design and use of games for learning.

What is motivation? *Motivation* is an internal state that initiates and maintains goal-directed behavior (Mayer, 2008, 2011). Let's explore each part of this definition: motivation is personal because it occurs within the learner; it is directed because it is aimed at accomplishing a goal; it is activating because it causes the learner to initiate action; and motivation is energizing because it causes the learner to persist and engage with intensity. Each of these aspects of the definition applies to someone who is intensely involved in playing a video game.

Table 3.9 summarizes the major motivational theories concerned with the question of what motivates people to exert effort in learning, and specifically, each is relevant to the question of what motivates people to exert effort in playing a game. Current theories of academic motivation cluster around four kinds of cognitions that learners bring with them to a learning experience: interests (as represented by interest theory and value theory); beliefs (as represented by self-efficacy theory, attribution theory, and self-theories); goals (as represented by goal orientation theories); and needs (as represented by self-determination theory and intrinsic motivation theory).

Interests: Interest Theory and Value Theory

Perhaps the most compelling reason that a video game is motivating is that it is fun to play. In other words, people are more likely to spend time playing a game that they enjoy, that interests them, or that they value. This rationale is most consistent with a classic theory of motivation, which can be called *interest theory*, as summarized in the first line of table 3.9. Interest

Table 3.9
Motivational theories related to games for learning

Theory	Description	Example
Interests		
Interest theory	People try harder to learn when they are interested in and value the material	I work hard because I like this game; it is interesting to me
Value theory	People try harder to learn when they find personal importance and value in what they are learning	I work hard because I value this game; it is important to me
Beliefs		
Self-efficacy theory	People try harder to learn when they see themselves as competent for the task	I work hard because I know I am good at this game; I am capable of doing this
Attribution theory	People try harder to learn when they attribute their successes and failures to effort rather than ability	I work hard on this game because I know my effort will pay off; if I fail, that means I need to try harder
Self-theories	People try harder when they believe intelligence is changeable rather than fixed	I can get smarter if I work hard to overcome challenges
Goals		
Goal orientation theory	People try harder to learn when they want to master the material	I work hard because I want to know everything there is to know about how this game works
Needs		
Self-determination theory	People try harder to learn when they believe they have control over the learning situation	I work hard because I am in charge of this game
Intrinsic motivation theory	People try harder to learn when they have internal rewards	I work hard because it makes me feel good to succeed in this game
Reinforcement theory	People try harder when they have been rewarded in the past	I work hard because I have been rewarded
Social cue theory	People try harder to learn when they have supportive social partners	I work hard because the game characters are working with me
Situated cognition theory	People try harder to learn when they can use their whole body	I work hard because I can move around as I play this game

theory proposes that people try harder to learn when they are interested in the material. A related cluster of motivational theories—called *expectancy-value theory* (or simply, *value theory*)—explores the similar idea that people try harder to learn when they value what they are doing (Wigfield, Tonks, & Klauda, 2009). In short, interest theory and value theory propose that people are more likely to give their best effort when they enjoy and value what they are doing, rather then when they find the learning task to be boring or pointless.

Interest theory has a long history in education and psychology, dating back to John Dewey's (1913) *Interest and Effort in Education*, in which he states: "Our policy of compulsory education rises or falls with our ability to make school life an interesting and absorbing experience" (p. ix). Although Dewey was not able to back up his theory with research evidence, more recent work has shown that students tend to learn more deeply when they study topics that interest them (Hidi, 2001; Pintrich, 2003; Schiefele, 1992; Schiefele, Krapp, & Winteler, 1992).

How does interest affect learning? One explanation is that interest creates general arousal, which facilitates learning, but Hidi and Baird (1986) have shown that this reasoning is inadequate. In contrast, a better-supported explanation is that interest causes learners to engage in appropriate cognitive processing during learning such as building cognitive connections with relevant prior knowledge. The implications for game design are that games should be designed in ways that are consistent with the game player's interests and values.

Can the addition of interesting elements spice up an otherwise-boring lesson to create what has been called *situational interest*? Dewey (1913) was adamant in his objections to situational interest as an instruction device: "When things have to make interesting, it is because interest itself is wanting. Moreover, the phrase is a misnomer. The thing, the object, is no more interesting than it was before" (p. 11–12). Although Dewey offered no evidence to back up his claim, more recent work shows that adding interesting but irrelevant facts to a paper-based text lesson (which can be called *seductive details*) detracts from learning (Garner, Gillingham, & White, 1989). Similarly, adding interesting but irrelevant facts or graphics (or even background music and sounds) to a computer-based multimedia lesson also detracts from learning (Harp & Mayer, 1997, 1998; Mayer, 2009). An important implication of research on seductive details is that the process of *gamification*—adding gamelike features to a traditional lesson—does not guarantee students will find the result to be an interesting and productive learning experience. This work suggests that game elements and instructional elements should be combined in a way that taps the interests and values that learners bring to the learning situation.

Beliefs: Self-Efficacy Theory, Attribution Theory, and Self-Theories

Learners have beliefs about themselves as learners that can affect the level of effort they are willing to exert on various learning tasks. Specifically, these beliefs can be about their competence for a learning task (as examined in self-efficacy theory), the role of their ability and effort in their success or failure in learning (as examined in attribution theory), and whether their ability is fixed or can be changed through learning (as examined in self-theories).

If you belief that you are capable of succeeding in a learning task (i.e., you have high self-efficacy for the task), you are likely to persist when things get difficult and to work intensely when needed. In contrast, if you believe you are not capable of succeeding in a learning task (i.e., you have low self-efficacy for the task), you are unlikely to persist when things get difficult, and rather than putting out the extra effort, you are likely to give up on mastering the material. These propositions summarize the basic tenets of self-efficacy theory, as summarized in the second row of table 3.9. *Self-efficacy* is a person's belief about their capability to perform successfully on a task (Schunk, 1991).

As research on self-efficacy reveals, students who have high self-efficacy for learning arithmetic actually do learn better than students who have low self-efficacy for it (Schunk, 1989; Schunk & Hanson, 1985). Similarly, students who believe they can do well in their first year of college obtain higher grades than students who do not believe they can do well (Chemers, Hu, & Garcia, 2001).

How does self-efficacy work? Schunk (1989) and Schunk and Pajares (2009) propose a model in which high self-efficacy (such as believing you will be good at a task) leads to stronger task engagement (such as persisting and trying hard), which in turn leads to higher achievement (such as better learning outcomes), which then leads to positive efficacy cues (such as good grades), thereby leading back to higher self-efficacy beliefs. In contrast, lower self-efficacy leads to weaker task engagement, which leads to lower achievement, which in turn leads to negative efficacy cues, thus leading back to lower self-efficacy beliefs.

Self-efficacy theory is relevant to games for learning when students who might otherwise have low self-efficacy for school-based learning tasks have high self-efficacy for certain kinds of game-based learning tasks. If academic content can be embedded in such games, students may persist and try hard to learn the academic material in games that they would not otherwise exert much effort for in school.

Next, consider a slightly different situation in which you receive an assessment of learning performance, such as a quiz grade, or being told whether or not you were correct on solving a math problem. If you got an A+ or solved the problem perfectly, you could attribute your performance to effort (e.g., "I really studied hard") or ability (e.g., "I am so smart"). If you failed the quiz or had the wrong answer on the problem, you also could attribute your performance to effort (e.g., "I really needed to study more) or ability (e.g., "I stink at this"). In short, learners may attribute their successes and failures to effort or ability (or a few other causes).

According to attribution theory (Graham & Williams, 2009; Weiner, 1992), students will be more likely to exert needed effort during learning if they attribute their successes and failures to effort rather than ability. For example, Borkowski, Weyhing, and Carr (1988), found that when they provided training aimed at helping students develop effort-based attributional beliefs, the students' learning performance improved.

Attribution theory is relevant to games for learning when students who might otherwise attribute their school successes and failures to ability come to believe that their game-based successes and failures depend on their level of effort. Thus, students may be more willing to persist in educational games than in traditional school learning venues.

Finally, consider beliefs about whether intelligence is fixed or changeable, which can be called self-theories (Dweck & Master, 2009). Consistent with attribution theory, if you think intelligence is fixed, you are more likely to attribute your failures to a lack of ability, and thereby experience helplessness; if you think intelligence is changeable, you are more likely to attribute failures to a lack of effort, and therefore use more effort-based strategies that result in performance improvement. Consistent with self-efficacy theory, if you think intelligence is fixed, your self-efficacy will decline as you experience failures; if you think intelligence is changeable, your self-efficacy will increase as you see that your effort can lead to improvements. Moreover, consistent with goal orientation theory, described below, if you think intelligence is fixed, you are more likely to want to prove your ability by looking good to your peers (i.e., have performance goals), and if you think intelligence is changeable, you are more likely to want to improve your ability (i.e., have mastery goals). In a recent review, Dweck and Master (2009) summarized research evidence that supports each of these points.

Self-theories are relevant to the design of educational games for two reasons. First, players' beliefs about intelligence can influence how they play the game, so that players with a belief in fixed intelligence may be less likely to respond productively to challenges and hence learn less from the game. Second, conversely, experience in game playing can affect players' beliefs about intelligence, so that success in figuring out how to overcome challenges in a game can help players develop the belief that intelligence is changeable. This shift in belief could even spill over into increased persistence on academic challenges on academic tasks outside of games. Games should be designed in ways that allow players to experience success on challenging tasks, thereby strengthening player's productive beliefs.

Goals: Goal-Orientation Theory

As shown in the third section of table 3.9, a major tenet of *goal-orientation theory* is that people bring goals to any learning experience that can affect their motivation to learn and in turn are affected by their learning experience (Maehr & Zusho, 2009). You have a *performance-approach goal* if you want to show others how well you perform, such as getting your name on the leader board. You have a *performance-avoidance goal* if you want to not look bad to others, such as not wanting to be the last one to reach the next level in a game you are playing. You have a *mastery goal* if you want to develop your competence on a task, such as figuring out everything there is know about the workings of a game you are playing. Maehr and Zusho (2009) summarize research evidence showing that having mastery goals is related to higher success in academic learning, whereas having performance-avoidance goals is related to lower success in academic learning. In some cases, performance-approach goals can be related to higher success in learning as well, but motivation theorists recommend that over the long run, people should be encouraged to develop mastery goals.

An important implication for game design is that players will be more likely to exert effort to learn deeply when a learning experience can prime mastery goals—that is, when the player's goal is to understand the material. Games should be designed in ways that make players want to achieve mastery for its own sake. There is not yet much evidence concerning whether certain learning experiences can help students develop mastery goals, but game playing offers an excellent venue for studying this intriguing idea. Players who start with performance-approach goals may eventually develop mastery goals as they spend more time with a game. The development of mastery goals with games could spill over into academic learning outside of games. These hypotheses form part of a potentially important research agenda on motivation to learn with computer games.

Needs: Self-Determination Theory and Intrinsic Motivation Theory

So far I have examined how people's interests, beliefs, and goals affect how hard they try to learn. The fourth element in table 3.9 concerns the learner's needs, including the need for autonomy and control, and even the need for reward.

Self-determination theory is the idea that people are born with an inherent motivation to learn, which is based on psychological needs for competence and autonomy (Ryan & Deci, 2009). *Intrinsic motivation* is based on internal mechanisms within the learner (such as the desire to learn), whereas *extrinsic motivation* is based on external rewards and punishments. Self-determination theory is consistent with research showing that offering external reward for intrinsically motivated activities can undermine intrinsic motivation (Lepper & Greene, 1978) and providing external punishment can also undermine intrinsic motivation (Ryan & Deci, 2009). Overall, Ryan and Deci (2009) summarize research showing that students work harder when teachers support student autonomy by allowing them some control over their learning experiences.

The implications for educational games are that players will try harder to learn when they have more control over the game, such as being able to personalize the game interface, determine the level of difficulty in the game, or even redo a challenging episode. Instead of emphasizing rewards and punishments as a means of creating extrinsic motivation, self-determination theory calls for designing games that assume players naturally want to master challenging situations, so successes and failures are simply information that the player uses on the way to becoming fully competent based on intrinsic motivation. A possible drawback is that forcing players to make choices about game mechanics may distract them from thinking about the to-be-learned material, so some balance is needed.

People try harder to learn when they are intrinsically motivated (i.e., driven by an internal desire to learn and the satisfaction that comes with mastery) rather than extrinsically motivated (i.e., controlled by external rewards and punishments). This is the basic tenet of intrinsic motivation theory. Malone and Lepper (1987; Lepper & Malone, 1987; Malone, 1981) offer a version of intrinsic motivation theory that is particularly suited to understanding the motivational power of games. According to Malone (1981), three elements make a game intrinsically motivational: challenge, fantasy, and curiosity. Concerning *challenge*, the game should require performance at a level that is slighter higher than the player's current level of competence, which can be achieved by building progressively more difficult levels into a game. Concerning *fantasy*, the game should allow the player to experience a sense of presence in an enticing environment that goes beyond the player's normal experience. Concerning *curiosity*, the game should reveal holes in the player's knowledge in a way that primes the player to want to make sense of the game. Although challenge, fantasy, and curiosity are sometimes touted as the defining features of games (as discussed in chapter 1), Malone does not offer compelling research evidence concerning how to implement them in ways that increase the effectiveness of games.

In a combined taxonomy of intrinsic motivation for learning, Lepper and Malone (1987; Malone & Lepper, 1987) concluded that intrinsic motivation is enhanced when instructional environments include challenge, fantasy, curiosity, and control. The new fourth element, *control*, refers to giving the learner a sense of control over the learning activity, consistent with similar recommendations from self-determination theory. Interpersonal elements that can create intrinsic motivation include cooperation, competition, and recognition. Although the taxonomy appears to have implications for educational games, research is needed to better clarify and test how to improve game effectiveness by priming intrinsic motivation in the learner. For example, does narrative theme help learning from games (as an attempt to create fantasy) or does adding a scoreboard help learning (as an attempt to create competition)? These kinds of questions are addressed in chapter 5 on the value-added approach to game research.

In stark contrast to self-determination theory and the intrinsic motivation theory from which it is derived, reinforcement theory (Skinner, 1938; Thorndike, 1911) is the idea that people are designed to be shaped by the external rewards and punishments they receive as a result of their actions, such as discussed in a previous section on theories of learning. If we consider reinforcement theory also as a theory of motivation, an important implication is that games have to provide right-wrong feedback so that players know when they made a correct or incorrect move. The classic version of reinforcement theory says that rewards (such as hearing applause when you make a shot) automatically strengthen a response and punishments (such as hearing a buzzer when you miss a shot) automatically weaken a response. In short, feedback in a game is reinforcement that automatically strengthens or weakens a response.

What is wrong with reinforcement theory as a theory of motivation? Although external rewards and punishments are effective in changing behavior, what is learned may not transfer to new situations, and may need continuing external rewards and punishments to be maintained (Mayer, 2009). However, research tends to show that when learners are intrinsically motivated to learn, they interpret feedback as information to guide their learning (Mayer, 2008). For example, higher-quality feedback that explains why an answer was right or wrong leads to better learning than simply telling players whether or not they were right (Hattie & Gan, 2011). As you can see, understanding the limitations of reinforcement theory has implications for the role of feedback in game design.

Two other emerging theoretical frameworks may also have implications for game design and therefore deserve mention: the need to learn in a social context and the need to learn in an embodied context. Social cue theory (Mayer, 2005) examines the cues in a computer-based multimedia learning environment that cause people to accept the computer as a social partner. For example, people try harder to learn when a lesson is presented in a human voice versus a machine voice, when the voice speaks in conversational style rather than a formal one, and when an on-screen agent uses humanlike gesture as opposed to no movement (Mayer, 2005; Mayer & DaPra, 2012). In a somewhat-broader approach, media equation theory (Reeves & Nass, 1996) looks at the conditions under which people will accept a computer as a social partner. The main implication for game design is that people will try harder to learn when they feel a game character is working with them. Chapter 5 explores whether people learn better from games when on-screen characters use friendly conversational style and thus provides a start on testing social cue theory.

Situated cognition theory (or embodiment theory) is based on the idea that cognition "depends not just on the brain but also on the body" (Robbins & Aydede, 2009, p. 3). For example, students remember a text about a farm better if they can physically manipulate toy objects referred to in the text such as a horse or barn during reading (Glenberg, Gutierrez, Levin, Japuntich, & Kaschak, 2004). The implications for game design are that players can learn better when they can move around as they play, such as tilting a tablet, touching a screen, or moving their arms and feet to control action in the game. Situated cognition theory (or embodiment theory) would predict that people learn better when a game takes place in immersive 3-D virtual reality (allowing the player to walk around in a simulated environment) rather than when the game is rendered on a desktop computer with keyboard input. Chapter 5 presents research evidence that disputes this prediction, however, so more work is needed in clarifying and testing the implications of situated cognition theory for game design.

Overall, there are many potentially useful links between theories of academic motivation and the design of games for learning, but there is a need for research that clarifies and tests specific research questions about how to prime motivation in games for learning.

Conclusion

The premise of this chapter is that the design of games for learning should be guided by an understanding of how the human mind works, including theories of learning and motivation. Concerning theories of learning, reinforcement theory is most relevant when the learning objective of a game is to help people acquire independent facts; schema theory is most relevant when the learning objective of a game is to help people build schemas; automaticity theory is most relevant when the learning objective of a game is to help people automatize their procedures; and social learning theory is most relevant when the learning objective of a game is to help people develop a collection of strategies.

Concerning theories of academic motivation, games are effective to the extent that they activate the most productive forms of the player's interests (as described in interest theory and value theory), beliefs (as described in self-efficacy theory, attribution theory, and self-theories), goals (as described in goal orientation theory), and needs (as described in self-determination theory and intrinsic motivation theory). Motivational theories, in essence, posit that game players will try harder to learn when they value, enjoy, and like the game; when they believe that they can succeed if they try hard, that successes and failures should be attributed to effort, and that their intelligence can be improved; when their goal is to master the material in the game; and when they feel that they have the autonomy and control to play the game as they wish.

The challenge of effective game design is to use game features that promote motivation to learn, but do not disrupt the appropriate cognitive processing during learning; and use instructional features that prime appropriate cognitive processes during learning, but do not shut down the player's motivation to learn. The theories of learning and motivation most relevant to game design offer some testable research hypotheses, which are examined next in chapters 4 through 7.

References

Anderson, L. W., Krathwohl, D. R., Airasian, P. W., Cruickshank, K. A., Mayer, R. E., Pintrich, P. R., et al. (2001). *A taxonomy for learning, teaching, and assessing: A revision of Bloom's taxonomy of educational objectives*. New York: Longman.

Baddeley, A. (1992). Working memory. *Science, 255*, 556–559.

Baddeley, A. (1999). *Human memory*. Boston: Allyn and Bacon.

Bandura, A. (1986). *Social foundations of thought and action: A social cognitive theory*. Hillsdale, NJ: Erlbaum.

Bartlett, F. C. (1932). *Remembering: A study in experimental and social psychology*. Cambridge: Cambridge University Press.

Borkowski, J. G., Weyhing, R. S., & Carr, M. (1988). Effects of attributional retraining on strategy-based reading comprehension in learning disabled students. *Journal of Educational Psychology, 80*, 46–53.

Chemers, M. M., Hu, L., & Garcia, B. F. (2001). Academic self-efficacy and first-year college student performance and adjustment. *Journal of Educational Psychology, 93*, 55–64.

Dewey, J. (1913). *Interest and effort in education.* Boston: Houghton Mifflin Company.

Dweck, C. S., & Master, A. (2009). Self-theories and motivation: Students' beliefs about intelligence. In K. R. Wentzel & A. Wigfield (Eds.), *Handbook of motivation at school* (pp. 123–140). New York: Routledge.

Ericsson, K. A., Charness, N., Feltovich, P. J., & Hoffman, R. R. (2006). *The Cambridge handbook of expertise and expert performance.* New York: Cambridge University Press.

Fitts, P. M., & Posner, M. I. (1967). *Human performance.* Belmont, CA: Brooks Cole.

Garner, R., Gillingham, M. G., & White, C. S. (1989). Effects of seductive details on macroprocessing and microprocessing in adults and children. *Cognition and Instruction, 6,* 41–67.

Glenberg, A. M., Gutierrez, T., Levin, J. R., Japuntich, S., & Kaschak, M. P. (2004). Activity and imagined activity can enhance young children's reading comprehension. *Journal of Educational Psychology, 96,* 424–436.

Graham, S., & Williams, C. (2009). An attributional approach to motivation in school. In K. R. Wentzel & A. Wigfield (Eds.), *Handbook of motivation at school* (pp. 11–34). New York: Routledge.

Harp, S. F., & Mayer, R. E. (1997). The role of interest in learning from scientific text and illustrations: On the distinction between emotional interest and cognitive interest. *Journal of Educational Psychology, 89,* 92–102.

Harp, S. F., & Mayer, R. E. (1998). How seductive details do their damage: A theory of cognitive interest in science learning. *Journal of Educational Psychology, 90,* 414–434.

Hattie, J. (2009). *Visible learning: A synthesis of over eight hundred meta-analyses relating to achievement.* New York: Routledge.

Hattie, J., & Gan, M. (2011). Instruction based on feedback. In R. E. Mayer & P. A. Alexander (Eds.), *Handbook of research on learning and instruction* (pp. 249–271). New York: Routledge.

Hidi, S. (2001). Interest, reading, and learning: Theoretical and practical considerations. *Educational Psychology Review, 13,* 191–209.

Hidi, S., & Baird, W. (1986). Interestingness: A neglected variable in discourse processing. *Cognitive Science, 10,* 179–194.

Johnson, C. I., & Mayer, R. E. (2010). Adding the self-explanation principle to multimedia learning in a computer-based game-like environment. *Computers in Human Behavior, 26,* 1246–1252.

Kapp, K. M. (2012). *The gamification of learning and instruction.* San Francisco: Pfeiffer.

Kirschner, P. A., Sweller, J., & Clark, R. E. (2006). Why minimal guidance during instruction does not work: An analysis of the failure of constructivist, discovery, problem-based, experiential, and inquiry-based teaching. *Educational Psychologist, 41*, 75–86.

Lepper, M. R., & Greene, D. (1978). *The hidden costs of reward*. Hillsdale, NJ: Erlbaum.

Lepper, M. R., & Malone, T. W. (1987). Intrinsic motivation and instructional effectiveness in computer-based education. In R. E. Snow & M. J. Farr (Eds.), *Aptitude, learning, and instruction: III. Conative and affective process analyses* (pp. 255–286). Hillsdale, NJ: Erlbaum.

Maehr, M. L., & Zusho, A. (2009). Achievement goal theory: The past, present, and future. In K. R. Wentzel & A. Wigfield (Eds.), *Handbook of motivation at school* (pp. 77–104). New York: Routledge.

Malone, T. W. (1981). Toward a theory of intrinsically motivating instruction. *Cognitive Science, 5*, 333–369.

Malone, T. W., & Lepper, M. R. (1987). Making learning fun: A taxonomy of intrinsic motivation for learning. In R. E. Snow & M. J. Farr (Eds.), *Aptitude, learning, and instruction: III. Conative and affective process analyses* (pp. 223–253). Hillsdale, NJ: Erlbaum.

Mayer, R. E. (2004). Should there be a three strikes rule against pure discovery? The case for guided methods of instruction. *American Psychologist, 59*(1), 14–19.

Mayer, R. E. (2005). Principles of multimedia learning based on social cues: Personalization, voice, and image principles. In R. E. Mayer (Ed.), *The Cambridge handbook of multimedia learning* (pp. 201–212). New York: Cambridge University Press.

Mayer, R. E. (2008). *Learning and instruction* (2nd ed.). Upper Saddle River, NJ: Pearson.

Mayer, R. E. (2009). *Multimedia learning* (2nd ed.). New York: Cambridge University Press.

Mayer, R. E. (2011). *Applying the science of learning*. Upper Saddle River, NJ: Pearson.

Mayer, R. E., & DaPra, S. (2012). An embodiment effect in computer-based learning with animated pedagogical agents. *Journal of Experimental Psychology: Applied, 18*, 239–252.

Mayer, R. E., & Johnson, C. I. (2010). Adding instructional features that promote learning in a game-like environment. *Journal of Educational Computing Research, 42*, 241–265.

Moreno, R., & Mayer, R. E. (1999). Multimedia-supported metaphors for meaning making in mathematics. *Cognition and Instruction, 17*, 215–248.

Paivio, A. (1986). *Mental representations: A dual coding approach.* Oxford: Oxford University Press.

Paivio, A. (2006). *Mind and its evolution: A dual coding theoretical approach.* Mahwah, NJ: Erlbaum.

Pellegrino, J. W., Chudowsky, N., & Glaser, R. (Eds.). (2001). *Knowing what students know: The science and design of educational assessment.* Washington, DC: National Academy Press.

Piaget, J. (1926). *The language and thought of the child.* London: Kegan, Paul, Trench, Trubner and Company.

Pintrich, P. (2003). Motivation and classroom learning. In W. M. Reynolds & G. E. Miller (Eds.), *Handbook of psychology: Vol. 7, educational psychology* (pp. 103–122). New York: Wiley.

Reeves, B., & Nass, C. (1996). *The media equation.* New York: Cambridge University Press.

Robbins, P., & Aydede, M. (2009). A short primer on situated cognition. In P. Robbins & M. Aydele (Eds.), *The Cambridge handbook of situated cognition* (pp. 2–10). New York: Cambridge University Press.

Ryan, R. M., & Deci, E. L. (2009). Promoting self-determined school engagement: Motivation, learning, and well being. In K. R. Wentzel & A. Wigfield (Eds.), *Handbook of motivation at school* (pp. 171–196). New York: Routledge.

Schiefele, U. (1992). Topic interest and levels of text comprehension. In K. A. Renninger, S. Hidi, & A. Krapp (Eds.), *The role of interest in learning and development* (pp. 151–182). Hillsdale, NJ: Erlbaum.

Schiefele, U., Krapp, A., & Winteler, A. (1992). Interest as a predictor of academic achievement: A meta-analysis of the research. In K. A. Renninger, S. Hidi, & A. Krapp (Eds.), *The role of interest in learning and development* (pp. 183–212). Hillsdale, NJ: Erlbaum.

Schunk, D. H. (1989). Self-efficacy and achievement behaviors. *Educational Psychology Review, 1,* 173–208.

Schunk, D. H. (1991). Self-efficacy and academic motivation. *Educational Psychologist, 26,* 207–231.

Schunk, D. H., & Hanson, A. R. (1985). Peer models: Influences on self-efficacy and achievement. *Journal of Educational Psychology, 77,* 313–322.

Schunk, D. H., & Pajares, F. (2009). Self-efficacy theory. In K. R. Wentzel & A. Wigfield (Eds.), *Handbook of motivation at school* (pp. 35–54). New York: Routledge.

Singley, M. K., & Anderson, J. R. (1989). *The transfer of cognitive skill.* Cambridge, MA: Harvard University Press.

Skinner, B. F. (1938). *The behavior of organisms: An experimental analysis.* Englewood Cliffs, NJ: Prentice Hall.

Skinner, B. F. (1968). *The technology of teaching.* Englewood Cliffs, NJ: Prentice Hall.

Sweller, J. (1999). *Instructional design in technical areas.* Camberwell, AU: ACER Press.

Sweller, J., Ayres, P., & Kalyuga, S. (2011). *Cognitive load theory.* New York: Springer.

Thorndike, E. L. (1911). *Animal learning.* New York: Hafner.

Thorndike, E. L. (1931). *Human learning.* New York: Century Company.

Weiner, B. (1992). Motivation. In M. Alkin (Ed.), *Encyclopedia of educational research* (6th ed., pp. 860–865). New York: Harper and Row.

Wigfield, A., Tonks, S., & Klauda, S. L. (2009). Expectancy-value theory. In K. R. Wentzel & A. Wigfield (Eds.), *Handbook of motivation at school* (pp. 55–76). New York: Routledge.

Wittrock, M. C. (1989). Generative processes of comprehension. *Educational Psychologist, 24,* 345–376.

II Evidence

4 Examples of Three Genres of Game Research

Chapter Outline

Summary

This chapter provides examples of how my colleagues and I have been trying to apply the science of learning to the three genres of game research described in chapter 1—value added, cognitive consequences, and media comparison. What does our research have to say about which game features are most effective in promoting academic learning? What does our research tell us about the cognitive consequences of playing off-the-shelf computer games? What does our research have to say about whether computer games are more effective in promoting academic learning than conventional media? This chapter provides examples of our attempts to answer these three questions using scientific research methodologies as described in chapter 2.

Three Genres of Game Research

As described in chapter 1, many strong claims are made for the educational value of computer games, but there is little strong empirical evidence to back up those claims.

In an early review of educational games, conducted over forty years ago, Abt (1970) observed, "The educational benefits of simulation games other than motivational are only dimly understood" (p. 18). More than three decades later, in a review of forty-eight empirical game studies, Hayes (2005) concluded, "The empirical research base on instructional effectiveness is fragmented, filled with ill-defined terms, and plagued with methodological flaws" (p. 3). Similarly, Perez (2008), noted that "there is a lack of empirical studies on the impact of games on learning and performance," and "the current research has little empirical guidance for the developers of educational games" (p. 287–288).

The solution to the need-for-evidence problem is to conduct scientifically rigorous studies of game effectiveness. As summarized in chapter 1, the consensus among scholars who have attempted to review research on the instructional effectiveness of games over the past twenty years is that the field needs a base of scientifically rigorous research evidence on what

Table 4.1
Three genres of game research

Research genre	Research question	Experimental comparison
Value added	Which features of a game promote learning?	Base versus enhanced version of game
Cognitive consequences	What do people learn from playing an off-the-shelf game?	Game versus no game
Media comparison	Do people learn better from a game or conventional media?	Game versus standard instruction

Source: Adapted from Mayer (2011).

works in game design (Connolly, Boyle, MacArthur, Hainey, & Boyle, 2012; Hayes, 2005; Honey & Hilton, 2011; O'Neil, Wainess, & Baker, 2005; Randel, Morris, Wetzel, & Whitehill, 1992; Sitzmann, 2011; Tobias, Fletcher, Dai, & Wind, 2011; Vogel et al., 2006; Young et al., 2012). The purpose of this chapter is to provide examples of how that needed evidence can be obtained in rigorous scientific experiments.

Table 4.1 shows how research on game effectiveness can be categorized into three genres: value-added research examines how adding various features in a game can affect learning, addressing the research question, "Which features improve the effectiveness of an educational computer game?"; cognitive consequences research examines what is learned from playing off-the-shelf computer games, addressing the research question, "What do people learn from playing a game?"; and media comparison research examines differences in learning the same content with a game versus conventional media, addressing the research question, "Do people learn better with games or conventional media?" In the next three sections, I present examples of game research in each of these genres that was carried out by my colleagues and me at the University of California, Santa Barbara (UCSB).

Examples of the Value-Added Approach

Figure 4.1 summarizes the value-added approach, which is intended to identify instructional features of a game that cause improvements in people's learning of the academic content of the game. In a value-added experiment, the primary independent variable is the presence or absence

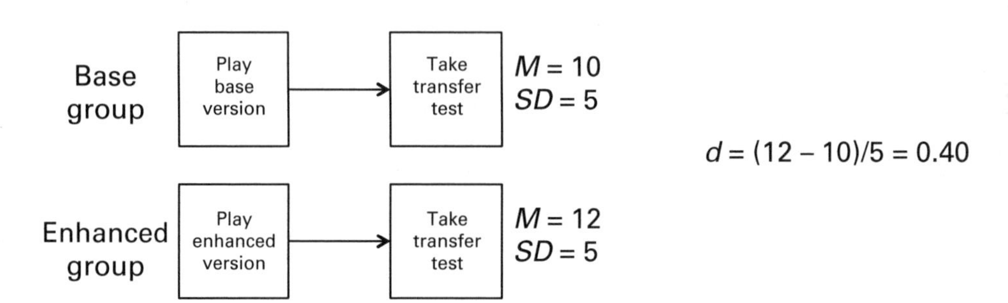

Figure 4.1
Value-added approach to game research

of an instructional feature (such as asking the learner to engage in self-explanation). As shown in figure 4.1, we randomly assign students to play a base version of the game (base group) or an enhanced version that involves adding an instructional feature to the base version (enhanced group). Next we give the same learning outcome test to both groups in order to determine what was learned. In a value-added experiment, the primary dependent variable is a measure of learning outcome such as how well learners perform on a problem-solving transfer test in which they must apply what they learned to a new situation. The main research question concerns whether people learn better when an instructional feature is added to a game. By using the mean score (*M*) and standard deviation (*SD*) for each group, we can calculate the effect size caused by the added feature (as described in chapter 2). An instructional feature attracts our attention when the effect size averages $d = 0.40$ or above over a large number of studies based on the same instructional feature.

In this section, I provide examples of the value-added research conducted by my colleagues and me involving five different computer-based games: *Design-a-Plant, Circuit Game, Profile Game, Virtual Factory*, and *Cache 17*. Each game was developed as a test bed for educational research rather than as a commercial product. Table 4.2 lists the instructional features we tested in each of the games using a value-added approach.

Table 4.2
Value-added research: Testing which features improve learning

Game	Features tested
Design-a-Plant	Modality, personalization, redundancy, image
Circuit Game	Explanative feedback, self-explanation, prompting, competition
Profile Game	Pretraining
Virtual Factory	Politeness
Cache 17	Narrative theme

Which Features Improve Learning in Design-a-Plant*?*

What Is the Design-a-Plant *Game?*

Consider the following scenario. You travel to another planet and meet Herman-the-Bug, who asks you to design a plant that will survive there. Herman-the-Bug describes the environmental conditions on the planet (such as low sunlight and little rainfall). You must choose the roots, stem, and leaves for your plant. Along the way, Herman-the-Bug offers feedback on your choices in which he includes short explanations of how plants grow. In all, you visit eight environments, such as the one shown on left side of figure 4.2, and you spend about twenty-five minutes on your adventures. This scenario describes a computer-based game called *Design-a-Plant* (Lester, Stone, & Stelling, 1999), which we used in a number of value-added experiments (Moreno, Mayer, Spires, & Lester, 2001; Moreno & Mayer, 2000, 2002a, 2002b, 2004).

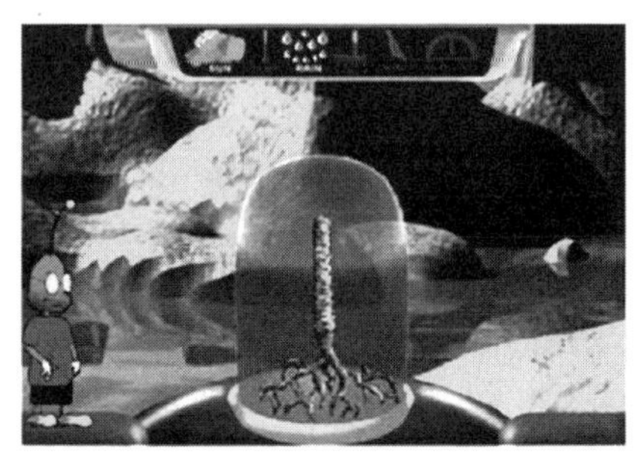

Base group: Play *Design-a-Plant* game. Herman-the-Bug communicates in printed text.

Enhanced group: Play *Design-a-Plant* game. Herman-the-Bug communicates in spoken form.

Transfer test: Determine best environment for various plants.

Figure 4.2
Do people learn better from playing *Design-a-Plant* when words are spoken rather than printed?

Modality Principle: Do People Learn Better from Playing Design-a-Plant *When the Words Are Spoken rather than Printed?*

Previous research in nongame environments suggests that people learn better in multimedia learning environments when words are spoken rather than printed, which can be called the modality principle (Mayer, 2009). The explanation for this modality principle is that when words are printed on the screen, learners must split their attention in the visual channel between the printed words and the graphic. For instance, when reading the on-screen text in the *Design-a-Plant* game, the learner may miss something going on in the animation, and when viewing the animation, the learner may miss something in the on-screen text. In contrast, when the words are spoken, learners can off-load the verbal material onto the verbal channel, thereby freeing up capacity in the visual channel for processing the animation.

In order to determine whether this modality effect also applies in game-like environments, we (Moreno et al., 2001; Moreno & Mayer, 2002a) compared student learning from the base version of *Design-a-Plant* (with printed words) versus with an enhanced version of *Design-a-Plant* (in which the words were spoken in a cartoonlike voice). As summarized in figure 4.2, in the base version of the game described above, Herman-the-Bug communicates via printed text on the bottom of the screen, whereas in the enhanced version Herman communicates via spoken words using an appealing cartoonlike voice. To test for what was learned from playing the game, we asked students to solve a series of transfer problems such as describing an appropriate environment (in terms of sunlight, wind, and rainfall) based on being given a picture of a plant with certain roots, stem, and leaves.

Table 4.3 summarizes the results of nine experimental comparisons involving both desktop platforms (Moreno et al., 2001; Moreno & Mayer, 2002a) and immersive virtual reality platforms (Moreno & Mayer, 2002a), in which students who played the enhanced version (with spoken words) performed better on the transfer test than did students who played the base version (with printed words), yielding a median effect size of $d = 1.41$, which is a large effect.

The set of findings summarized in table 4.3 provides consistent evidence for the *modality principle of game design*: people learn better in multimedia games where instructional words are spoken as opposed to printed. Research in nongame multimedia learning environments, however, suggests that

Table 4.3
Modality principle: People learn better in games when words are spoken rather than printed

Source	Effect size
Moreno et al. (2001, expt. 4a)	0.60
Moreno et al. (2001, expt. 4b)	1.58
Moreno et al. (2001, expt. 5a)	1.41
Moreno et al. (2001, expt. 5b)	1.71
Moreno & Mayer (2002a, expt. 1a)	0.93
Moreno & Mayer (2002a, expt. 1b)	0.62
Moreno & Mayer (2002a, expt. 1c)	2.79
Moreno & Mayer (2002a, expt. 2a)	0.74
Moreno & Mayer (2002a, expt. 2b)	2.24
Median	1.41

there may be important boundary conditions for the modality principle, in which the principle applies most strongly when the material is complex, the presentation is fast paced, and the learners are familiar with the words (Mayer, 2009).

Personalization Principle: Do People Learn Better in Design-a-Plant *When the Words Are in a Conversational versus Formal Style?*

Research on multimedia learning in nongame environments offers support for the personalization principle: people learn better when words are presented in conversational style rather than formal style (Mayer, 2009). The rationale for using conversational style is that people try harder to make sense out of a message when they feel they are in a conversation with a social partner—a situation that is primed by using a conversational style (Mayer, 2009; Nass & Brave, 2005; Reeves & Nass, 1996).

Let's examine what happens when we apply this principle to the conversational style reflected in Herman-the-Bug's words in *Design-a-Plant* (Moreno & Mayer, 2000, 2004), as summarized in figure 4.3. In the base version of the game, Herman-the-Bug uses formal conversational style. For example, in explaining plant growth in rainy environments, he says, "In very rainy environments, plant leaves have to be flexible so they are not damaged by the rainfall. What really matters for the rain is the choice between thick and thin leaves."

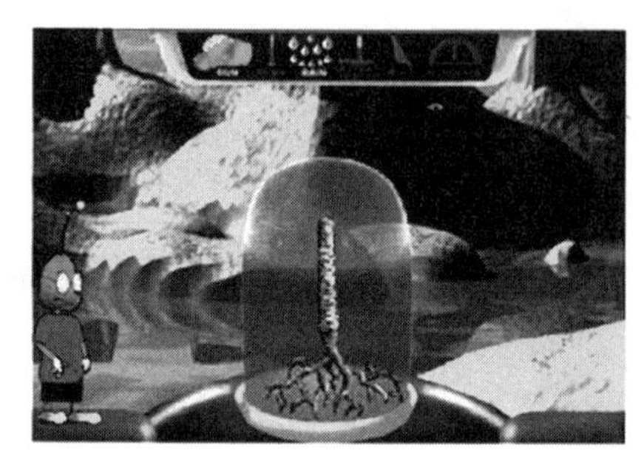

Base group: Play *Design-a-Plant* game. Herman-the-Bug communicates using formal style (e.g., "This program is about what type of plant survives on different planets").

Enhanced group: Play *Design-a-Plant* game. Herman-the-Bug communicates in conversational style (e.g., "You are about to begin a journey, where you will be visiting different planets").

Transfer test: Determine best environment for various plants.

Figure 4.3
Do people learn better from playing *Design-a-Plant* when words are in a conversational versus formal style?

In the enhanced version of the game, Herman communicates in a personalized conversational style in which he uses first- and second-person constructions (e.g., "I," "you," and "your"), thereby speaking directly to the player. For instance, in explaining plant growth in rainy environments, he says, "This is a very rainy environment and the leaves of your plant have to be flexible so they're not damaged by the rainfall. What really matters for the rain is your choice between thick and thin leaves."

Table 4.4 summarizes the results of five separate experiments involving both spoken formats (Moreno & Mayer, 2000, 2004) and printed formats (Moreno & Mayer, 2000), in which we compared students' transfer test performance after playing the base version of the game (with a formal style) versus the enhanced version (with a personalized style). In all five comparisons, players performed better on a transfer test if the words were in a personalized conversational rather than a formal style, yielding a median effect size of $d = 1.58$, which is a large effect. This pattern is consistent with research in nongame multimedia learning environments (Mayer, 2009), and allows us to offer the *personalization principle of game design*: people learn better from games in which words are in a personalized conversational style rather than in a formal style. Possible boundary conditions include that the principle may be strongest when the players are not familiar with the game environment and when personalization is not overdone (Mayer, 2009).

Table 4.4
Personalization principle: People learn better in games when words are in conversational rather than formal style

Source	Effect size
Moreno & Mayer (2000, expt. 3)	1.92
Moreno & Mayer (2000, expt. 4)	1.49
Moreno & Mayer (2000, expt. 5)	1.11
Moreno & Mayer (2004, expt. 1a)	1.58
Moreno & Mayer (2004, expt. 1b)	1.93
Median	1.58

Redundancy Principle: Do People Learn Better in Games When the Words Are Printed and Spoken Rather than Spoken Alone?

In redesigning the *Design-a-Plant* game, you might be tempted to include both spoken text (in Herman's friendly voice) and the corresponding printed text (perhaps at the bottom of the screen) in order to accommodate a variety of learning styles. However, previous research on multimedia learning in nongame environments has yielded the redundancy principle: people learn better from multimedia lessons when words are spoken rather than when words are spoken and printed. The theoretical rationale is that redundant spoken and printed text can prime extraneous processing in which learners waste their limited processing capacity on reconciling the two incoming verbal streams. In addition, presenting simultaneous captions and narration creates split visual attention in which people may miss some of the animation while reading the caption, or they may miss some of the caption while viewing the animation.

What would happen if Herman communicated in both printed and spoken text in *Design-a-Plant* so the player presumably could have the choice of reading printed words or listening to spoken words? As shown in figure 4.4, we compared the transfer test performance of students who received a spoken version of *Design-a-Plant,* in which Herman's words are in spoken form to a spoken-plus-printed version in which Herman's words are presented simultaneously in both formats (Moreno & Mayer, 2002b).

Table 4.5 shows that on a subsequent transfer test, we found a small difference favoring the spoken version, yielding a median effect size of $d = -0.22$ across two experimental comparisons. Unlike research in nongame multimedia learning environments (Mayer, 2009) in which adding redundant text greatly hurt learning, we found simply that adding

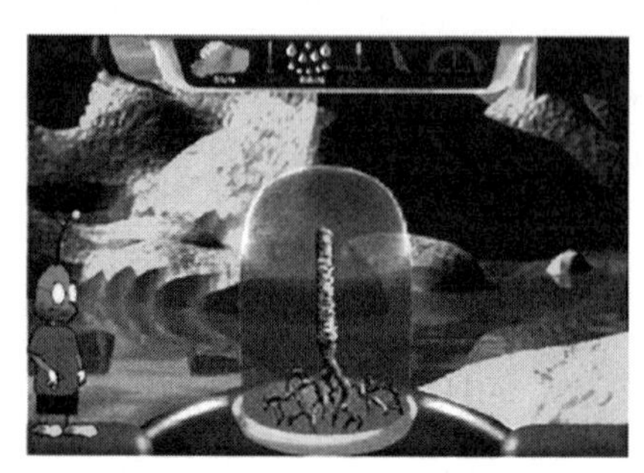

Base group: Play *Design-a-Plant* game. Herman-the-Bug communicates in spoken text.

Enhanced group: Play *Design-a-Plant* game. Herman-the-Bug communicates in spoken form and with identical on-screen printed text.

Transfer test: Determine best environment for various plants.

Figure 4.4
Do people learn better from playing *Design-a-Plant* when words are printed and spoken rather than spoken only?

Table 4.5
Redundancy principle: People do not learn better in games when words are printed and spoken rather than spoken alone

Source	Effect size
Moreno & Mayer (2002b, expt. 2a)	–0.19
Moreno & Mayer (2002b, expt. 2b)	–0.25
Median	–0.22

redundant text did not help learning from games. Based on the current level of research evidence, we can offer only this modest version of the *redundancy principle of game design*: people do not learn better from games that present redundant printed and spoken words rather than spoken words alone. These preliminary findings do not appear to supply an evidence-based rationale for adding on-screen printed text to spoken text in educational games. Some possible boundary conditions may be that on-screen words are helpful when the material is technical, the message is complicated, or the players are not native speakers of the language (Mayer, 2009).

Image Principle: Do People Learn Better from Playing Design-a-Plant *When the Agent's Image Is on the Screen?*

So far, we have focused on the on-screen agent's words in the *Design-a-Plant* game. Let's shift for a moment to the role of the agent's image on the screen. In multimedia research in nongame environments, there is preliminary evidence that including a static image of a pedagogical agent on

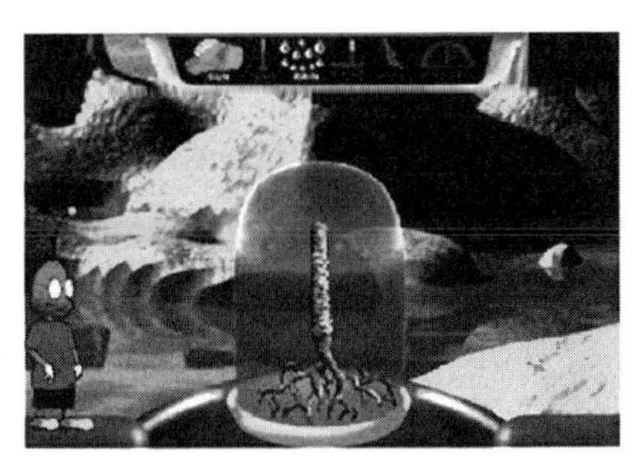

Base group: Play *Design-a-Plant* game. Herman-the-Bug's image is not on the screen.

Enhanced group: Play *Design-a-Plant* game. Herman-the-Bug's image is on screen.

Transfer test: Determine best environment for various plants.

Figure 4.5
Do people learn better from playing *Design-a-Plant* when the agent's image is on the screen?

the screen does not improve student learning (Mayer, 2009; Mayer & DaPra, 2012). This work suggests the image principle: people do not learn better from multimedia lessons when the agent's image is on the screen.

In order to determine whether the image principle applies to *Design-a-Plant*, we compared the transfer test performance of students who played the game with or without Herman's image on the screen, as summarized in figure 4.5. In the enhanced version of the game, Herman's image is on the screen when he is talking. Players see a cartoonlike character who is communicating with them. In the base version, we removed Herman's image from the screen, so that all that is left is his voice (or his printed words on the screen).

Table 4.6 shows that across four experimental comparisons (Moreno et al., 2001), retaining the agent's image did not have a strong effect on transfer test performance, yielding a median effect size of $d = 0.22$, which is

Table 4.6
Image principle: People do not learn better in games when the agent's image is on the screen

Source	Effect size
Moreno et al. (2001, expt. 4a)	–0.50
Moreno et al. (2001, expt. 4b)	0.22
Moreno et al. (2001, expt. 5a)	0.22
Moreno et al. (2001, expt. 5b)	0.35
Median	0.22

a small effect favoring the image version over the no-image version. Based on these findings, we can offer the *image principle of game design*: people do not learn much better from games in which an agent's image is on the screen. It is possible that the voice or text carries enough social cues to allow the learner to form a social partnership, or perhaps the agent needs to behave in a more lifelike way to have a strong effect on learning.

Overall, the value-added approach to research on *Design-a-Plant* yields two features that strongly improve learning—modality and personalization—and two features that do not—redundancy and image. As you can see, the value-added approach offers a technique for identifying features that have the greatest promise for increasing the instructional effectiveness of *Design-a-Plant*.

Which Features Improve Learning in the Circuit Game*?*

What Is the Circuit Game*?*

The *Circuit Game* is a puzzle game intended to help players learn how electric circuits work, consisting of ten levels with the tenth level serving as an embedded transfer test (Johnson & Mayer, 2010; Mayer & Johnson, 2010). As exemplified on the left side of figure 4.6, each level consists of a series of

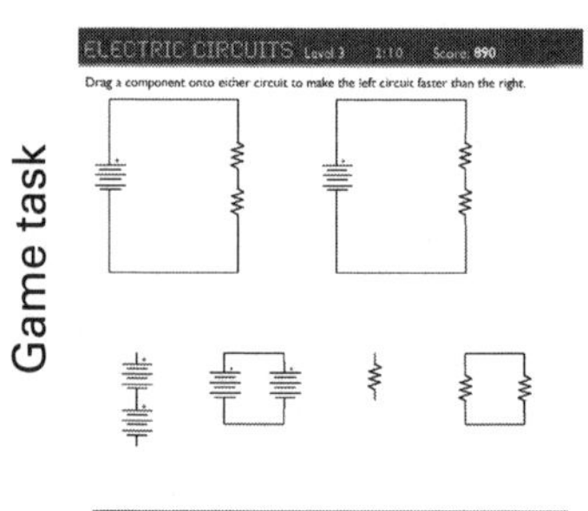

Base group: Play *Circuit Game* without explanative feedback.

Enhanced group: Play *Circuit Game* with explanative feedback after each key move.

Transfer test: Solve 25 new circuit problems on level 10 as an embedded test.

Feedback

ELECTRIC CIRCUITS Level 1 0:06 Score: -10

Which circuit is faster?

SAME

Figure 4.6

Do people learn better from playing the *Circuit Game* when they get explanative feedback after key moves?

circuit problems such as determining which of two circuits has a higher rate of current flow, or making two circuits have the same current by dragging and dropping an appropriate component into one of them. If the player makes a correct move, the computer makes a beep and adds fifty points to the player's score (indicated on the top of the screen); if the player makes an incorrect move, the computer makes a buzzer sound and ten points are deducted from the player's score. In the tenth level, the player receives an embedded transfer test consisting of a series of twenty-five problems that involve judging which of two lights will burn brighter in a number of different circuits. Unlike levels one through nine, there is no feedback, and the problems involve actual lightbulbs in circuits. The game takes about thirty minutes to complete.

Explanative Feedback Principle: Do People Learn Better from Playing the Circuit Game *When They Get Explanative Feedback after Key Moves?*

Explanative feedback occurs when someone performs a task and receives an explanation about the quality of the performance. Previous research in educational psychology suggests that explanative feedback is one of the most powerful techniques for improving learning in nongame environments (Hattie, 2009; Hattie & Gan, 2011; Shute, 2008), yielding what could be called the explanative feedback principle: people learn better when they get explanative feedback on their performance. The theoretical rationale for explanative feedback is that it encourages learners to process the material more deeply—that is, it fosters generative processing.

Does the explanative feedback principle apply to learning with games such as the *Circuit Game*? Figure 4.6 summarizes an experimental comparison between a base group that plays the *Circuit Game* without explanative feedback and an enhanced group that plays with explanative feedback. As you can see, the base version of the *Circuit Game* described above provides right-wrong feedback (by including a beep or buzzer in response to the player's moves), but no explanation is given. In contrast, in the enhanced version of the *Circuit Game*, when someone solves a circuit problem, a dialogue box appears on the screen showing the correct answer along with the relevant rule (such as, "If you add a battery in serial, the flow rate increases.").

Table 4.7 shows the results of one experimental comparison on explanative feedback (Mayer & Johnson, 2010), in which students perform much better on the transfer test when the first nine levels of the game include

Table 4.7
Explanative feedback principle: People learn better in games when they get explanative feedback after key moves

Source	Effect size
Mayer & Johnson (2010)	0.68

explanative feedback (d = 0.68). This finding is consistent with the nongame literature and offers preliminary evidence in support of the *explanative feedback principle for games*: people learn better in games when they receive explanative feedback after key moves.

Self-Explanation Principle: Do People Learn Better from Playing the Circuit Game *When They Are Asked to Select an Explanation for Their Moves?*
Self-explanation is an active learning strategy in which learners explain confusing material to themselves during learning. In previous research involving learning in nongaming environments, students learned more deeply when they were encouraged to engage in self-explanation—that is, when they are asked to talk aloud as they read a lesson in an effort to explain the material to themselves (Roy & Chi, 2005). Overall, previous research in educational psychology suggests that self-explanation is one of the most powerful techniques for improving learning in nongame environments (Fonseca & Chi, 2011; Roy & Chi, 2005), yielding what could be called the self-explanation principle: people learn better when they are prompted to provide explanations during learning (Roy & Chi, 2005). The theoretical rationale for the self-explanation principle is that the act of self-explaining encourages learners to process the material more deeply—that is, it fosters generative processing.

Figure 4.7 summarizes research aimed at testing the self-explanation principle within the context of the *Circuit Game*, in which we compare a base version of the game with an enhanced version that includes self-explanation prompts (Mayer & Johnson, 2010). As you can see, the base version of the game described in the previous section does not provide much opportunity for learners to reflect on what they are doing. In an attempt to encourage learners to think more deeply, we created an enhanced version of the *Circuit Game* that involved a form of self-explanation in which players were asked to click on a rationale for their move from a list

Game task

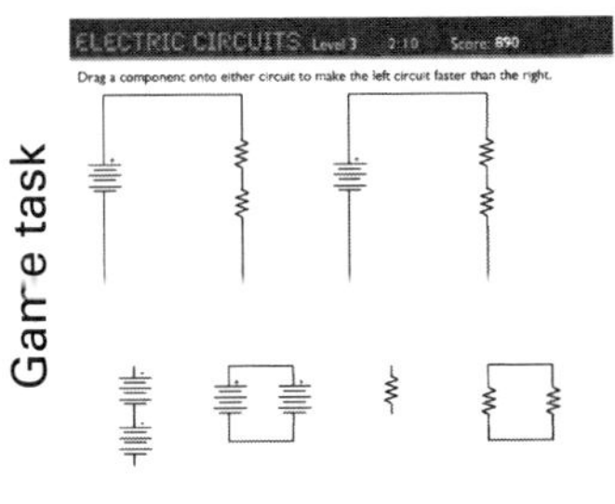

Base group: Play *Circuit Game* without self-explanation.

Enhanced group: Play *Circuit Game* with self-explanation.

Transfer test: Solve 25 new circuit problems on level 10 as an embedded test.

Self-explanation

Figure 4.7
Do people learn better from playing the *Circuit Game* when they are asked to select an explanation for their moves?

for each problem in levels one through nine. After each move, a dialogue box appeared with a list of possible reasons for the move, and the player's job was to click on the reason for the correct answer. This modified version of self-explanation was included to minimize disruption to game flow while encouraging players to think about the reason for each move they made.

Table 4.8 summarizes the results of three experimental comparisons, in which we compared the transfer test performance of low-knowledge college students who played the base or enhanced version of the *Circuit Game* (Johnson & Mayer, 2010; Mayer & Johnson, 2010). Students who played the enhanced version performed better on a transfer test than did students

Table 4.8
Self-explanation principle: People learn better in games when they are asked to select an explanation for their moves

Source	Effect size
Mayer & Johnson (2010)	0.90
Johnson & Mayer (2010, expt. 1)	1.20
Johnson & Mayer (2010, expt. 2)	0.71
Median	0.90

who played the base version, yielding a median effect size of $d = 0.90$, which is considered a large effect. This research shows that the self-explanation effect found in nongame environments (Roy & Chi, 2005) extends to game-like environments such as the *Circuit Game*. This research offers preliminary evidence for the *self-explanation principle in computer games*: people learn better from games when they are asked to explain their actions. Interestingly, Johnson and Mayer (2010) found that asking students to type in their reasons for each key move in an on-screen text box did not improve learning in the *Circuit Game*, perhaps because it is highly disruptive to game play. Another possible boundary condition is that the students in this study lacked domain knowledge, so it is possible that the principle may apply mainly to low-knowledge players.

Prompting Principle: Do People Learn Better from Playing the Circuit Game *When They Are Prompted to Reflect on the Underlying Principles?*
A serious limitation of games for learning is that players may become so involved in game action that they do not have time to reflect on what they are learning. What would happen if players were explicitly asked to reflect on the principles underlying game play throughout the game? As you can see, prompting is closely related to self-explanation because both are intended to prime active learning strategies aimed at sense making. The prompting principle is that people learn more deeply when they are asked to reflect on their learning. The theoretical explanation for the prompting principle is that explicitly prompting students to reflect during game play can encourage them to process the material more deeply—that is, to engage in generative processing.

Figure 4.8 summarizes research aimed at testing the prompting principle within the context of the *Circuit Game*, in which we compare a base version of the game with an enhanced version that includes prompts to relate the material to the underlying principles of electricity (Fiorella & Mayer, 2012). The base version of the game described in the previous section does not provide much opportunity for learners to reflect on what they are doing. In an attempt to encourage learners to think more deeply, we created an enhanced version of the *Circuit Game* that included prompting in the form of asking players to complete a worksheet during the game in which they select the correct version of each of eight electricity principles, such as "If you add a battery in serial, the flow rate (circle one): increases, decreases, or

Game task

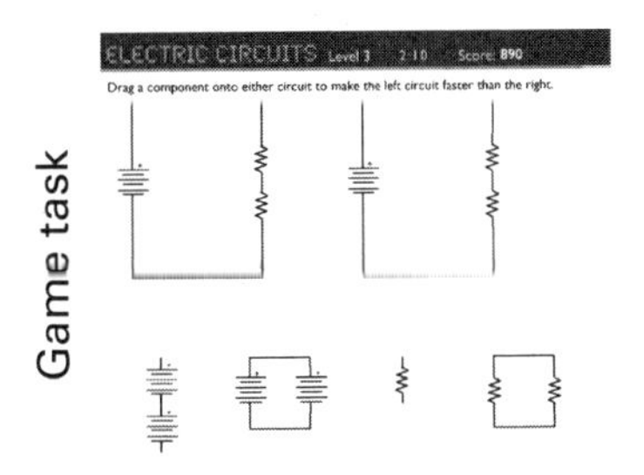

Base group: Play *Circuit Game* without prompts.

Enhanced group: Play *Circuit Game* with prompts.

Transfer test: Solve 25 new circuit problems on level 10 as an embedded test.

Prompts

Number	Principle
1	If you add a battery in serial, the flow rate (circle one): increases, decreases, or does not change
2	If you add a battery in parallel, the flow rate (circle one): increases, decreases, or does not change.
3	If you add a resistor in serial, the flow rate (circle one): increases, decreases, or does not change.
4	If you add a resistor in parallel, the flow rate (circle one): increases, decreases, or does not change.
5	If you take away a battery in serial, the flow rate (circle one): increases, decreases, or does not change.
6	If you take away a battery in parallel, the flow rate (circle one): increases, decreases, or does not change.
7	If you take away a resistor in serial, the flow rate (circle one): increases, decreases, or does not change.
8	If you take away a resistor in parallel, the flow rate (circle one): increases, decreases, or does not change.

Figure 4.8
Do people learn better from playing the *Circuit Game* when they are prompted to reflect on the underlying principles?

does not change." In another version of prompting, we supplied a worksheet containing the eight electricity principles and asked players to determine which one applied to each of their key choices in the game. We also asked them to say how many batteries and resistors were in each circuit, and whether they were in series or parallel.

Table 4.9 shows that in two out of two experimental comparisons, students performed better on a transfer test when they played the enhanced version of the *Circuit Game* rather than the base version, yielding a median effect size of d = 0.65, thus favoring the inclusion of prompting. These

Table 4.9
Prompting principle: People learn better in games when they are prompted to reflect on the underlying principles

Source	Effect size
Fiorella & Mayer (2012, expt. 1)	0.77
Fiorella & Mayer (2012, expt. 2)	0.53
Median	0.65

results support the *prompting principle for games*: people learn better in games when they are prompted to reflect on the underlying principles. An important boundary condition detected by Fiorella and Mayer (2012) is that players who were not able to complete the worksheet correctly did not benefit, suggesting that low-performing learners may need more structured prompts.

Competition Principle: Do People Learn Better from Playing the Circuit Game *When Competitive Features Are Added?*

Another way to spice up a lesson in hopes of increasing player engagement is to add competitive features to a game. Rieber (2005) reported that children prefer games that have competitive elements, and Clark, Lawrence, Astley-Jones, and Gray (2009) observed that the reward centers of the brain are activated when a game involves competition. Thus, the theoretical rationale for adding competitive features is that they are intended to increase the player's intrinsic motivation to engage and persist in game activity (Lepper & Malone, 1987; Malone, 1981; Malone & Lepper, 1987).

Figure 4.9 summarizes an experimental comparison between a group that receives the base version of the *Circuit Game* with one that receives an enhanced version with competitive features added (DeLeeuw & Mayer,

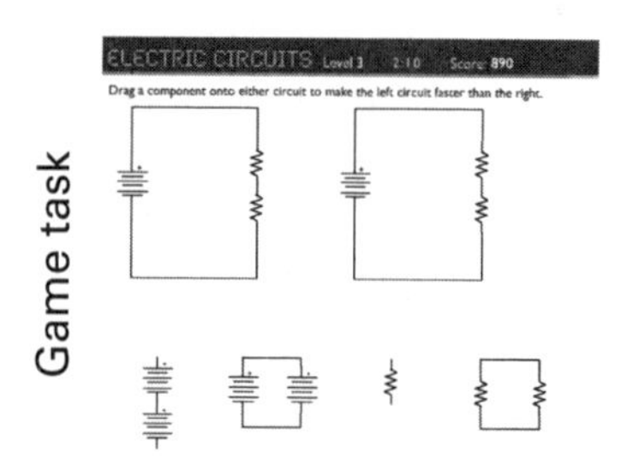

Figure 4.9
Do people learn better from playing the *Circuit Game* when competitive features are added?

Table 4.10
Competition principle: Women learn better and men learn worse when competition is added to a game, but there is no overall effect

Source	Effect size
DeLeeuw & Mayer (2011, women)	0.24
DeLeeuw & Mayer (2011, men)	–0.54
DeLeeuw & Mayer (2011, all)	–0.06

2011). In the competitive version of the game, players earned up to ten tickets per level that were then placed in a $50 raffle. An on-screen scoreboard for each level kept a running count of how many tickets the player had won (or lost) so far.

Table 4.10 shows that although DeLeeuw and Mayer (2011) found no overall effect of competition ($d = –0.06$ favoring the base version), there was a pattern in which women's transfer test performance was improved by the addition of competitive features ($d = 0.24$), but men's transfer test performance was hurt by the addition of competitive features ($d = –0.54$). These preliminary results do not offer strong support for including competitive factors in games, possibly because they may cause some players (particularly men and boys) to focus more on game activity than on reflection. The *competition principle for games* is that overall, people do not learn better when competitive features are added to games.

Value-added research on the *Circuit Game* yielded three promising instructional features for consideration for inclusion in computer games—explanative feedback, self-explanation, and prompting—and one somewhat-unpromising feature that was not supported in our preliminary research—competition.

Which Features Improve Learning in the Profile Game*?*

What Is the Profile Game*?*

The *Profile Game* is a computer-based geology simulation game in which the player surveys an area of a planet's surface in order to identify the presence of a hidden geologic feature, such as a basin, island, ridge, seamount, or trench (Prothero, 1997). As shown in figure 4.10, players can click on any two points in the exploration window shown on the top of the screen and they will see a profile indicating the distance from sea level for each point

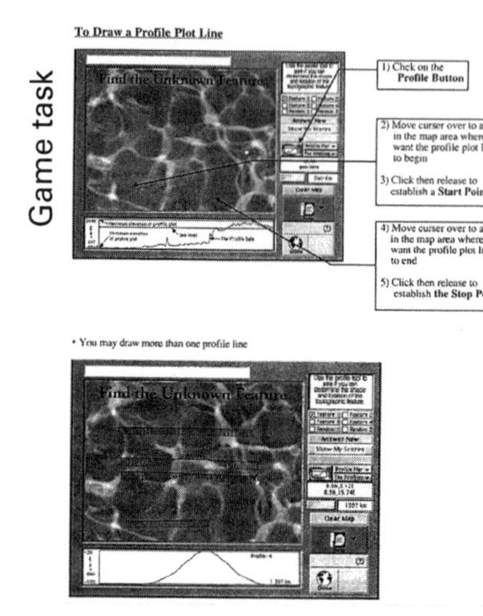

Base group: Play *Profile Game* without pretraining.

Enhanced group: Play *Profile Game* with pretraining.

Transfer test: Solve new profile problems.

Figure 4.10
Do people learn better from playing the *Profile Game* when they receive pretraining in the key concepts?

on the line. Players can draw as many lines as they wish, until they think they know which geologic feature is present. At that point, they give their answer by clicking on the appropriate boxes at the bottom of the screen indicating basin, island, ridge, seamount, or trench, and whether the feature is on land or in the ocean. In a typical game session, players work on five problems and spend about ten minutes. The game is intended to promote authentic scientific reasoning in geology.

Pretraining Principle: Do People Learn Better from the Profile Game *When They Receive Pretraining in the Key Concepts?*
Research on multimedia learning in nongame environments has suggested the pretraining principle: people learn better from a multimedia message when they receive pretraining in the names and characteristics of the key concepts (Mayer, 2009). The theoretical rationale for the pretraining principle is that pretraining in key concepts allows people to use their limited cognitive resources to build connections among concepts during learning rather than having to figure out what the key concepts are.

Figure 4.10 summarizes an experimental comparison between the base version of the *Profile Game* and an enhanced version in which a pretraining sheet is added. As you can see, the base version offers a challenge to students who are not familiar with surveying strategies in geology. Players are given instructions and then solve five problems. However, the task of learning surveying strategies may be somewhat hampered if players do not have appropriate schemas for the various geologic features—that is, pictorial representations of each of the possible features. Game playing in the base version may suffer because players have to build pictorial schemas as well as develop effective surveying strategies.

To address this problem, in the enhanced group we added a brief pretraining period in which we gave learners a pretraining sheet that showed drawings of each of the possible geologic features—ridge on land, ridge in the ocean, trench on land, trench in the ocean, basin on land, basin in the ocean, island in the ocean, and seamount in the ocean. The goal was to help learners be able to visualize the geologic features in the game.

In order to determine the effectiveness of adding pretraining, we compared the problem-solving performance of inexperienced college students who played the original version of the game with those who played the same game with pretraining (Mayer, Mautone, & Prothero, 2002). As a transfer test, students were given a series of five sheets of paper, each showing the exploration window for a new problem with five to nine profile lines along with a space for them to indicate the type of geologic feature.

Table 4.11 shows that across two experiments (Mayer, Mautone, & Prothero, 2002), the pretraining group solved more problems correctly than did the non-pretraining group, yielding a median effect size of d = 0.71, which is in the medium-to-large range. These results provide preliminary evidence for the *pretraining principle of game design*: players learn better when they are given pretraining in the names and characteristics of the key concepts in the game. A possible boundary condition may be that

Table 4.11

Pretraining principle: People learn better in games when they receive pretraining in key concepts

Source	Effect size
Mayer, Mautone, & Prothero (2002, expt. 2)	0.57
Mayer, Mautone, & Prothero (2002, expt. 3)	0.85
Median	0.71

pretraining is especially effective for players who have low levels of experience in the content domain of the game, as was the case in the *Profile Game* experiment (Mayer, Mautone, & Prothero, 2002).

Which Features Improve Learning in the Virtual Factory Game*?*

What Is the Virtual Factory Game*?*

The *Virtual Factory Game* is a Web-based engineering simulation game in which learners are asked to design an assembly line for a product based on analyzing the history of previous orders (Dessouky, Verma, Bailey, & Rickel, 2001). In the game, students are given an introduction to a factory problem, and must forecast future product demand, develop a production plan, and schedule the processing of jobs. As shown on left side of figure 4.11, students work through a computer interface in which they can enter actions and view the current state of the factory in a window on the upper left side of the screen, chat with an on-screen agent on the upper right side of the screen, and receive tutorial information on the lower right side of the screen. Students need about thirty-five minutes to complete the task.

Politeness Principle: Do People Learn Better from Playing Virtual Factory *When the On-screen Agent Uses Polite Rather than Direct Wording?*

The politeness principle is that people learn better when instructors give feedback and offer suggestions using polite rather than direct wording (Wang et al., 2008). The theoretical rationale proposed by Brown and Levinson (1987) is that direct statements telling the learner what to do pose

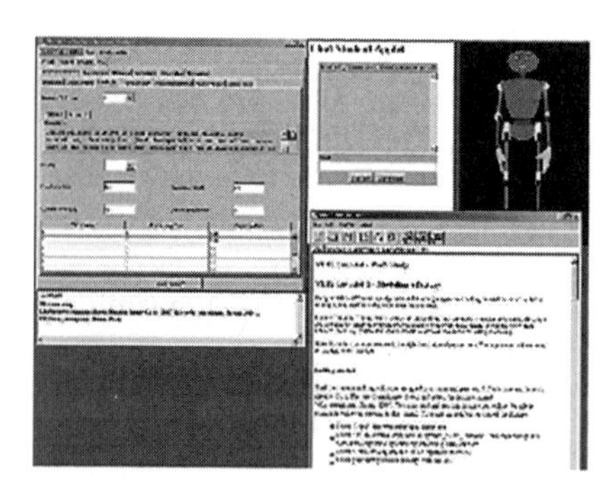

Base group: Play *Virtual Factory* with on-screen agent who uses direct wording (e.g., "Press the ENTER key").

Enhanced group: Play *Virtual Factory* with on-screen agent who uses polite wording (e.g., "Shall we press the ENTER key?").

Transfer test: Answer comprehension questions.

Figure 4.11
Do people learn better from playing the *Virtual Factory Game* when the on-screen agent uses polite wording rather than direct wording?

threats to negative face (by not allowing the learner to have freedom of action) and a threat to positive face (by not working with the learner). When people feel threatened, they are less likely to work effectively with an instructor.

In *Virtual Factory*, the on-screen agent offers feedback and guidance in response to the learner's actions and questions. Figure 4.11 summarizes an experimental comparison between a base group that plays a version of *Virtual Factory* in which the on-screen agent uses direct wording and an enhanced group that plays a version of *Virtual Factory* in which the on-screen agent uses polite wording. In the base version, the on-screen tutor uses a direct conversational style in which he simply tells the learner what to do when the learner asks for help or gets stuck. For example, a direct comment is: "Read the paragraph more carefully."

In contrast, we also designed a polite version of the game that was identical to the original one except the tutor used a polite conversational style for feedback and guidance instead of a direct conversational style. For instance, more polite ways to say the same thing are: "Why don't we read the paragraph more carefully?" or "You might want to read the paragraph." The wording of the polite comments was based on Brown and Levinson's (1987) politeness theory, which specifies wordings designed to save negative face (i.e., the desire to be unimpeded by others) and save positive face (i.e., the desire to be respected by others). In the polite version of *Virtual Factory*, each direct statement was reworded into polite form to reduce threats to positive and negative face.

We asked college students who were unfamiliar with engineering to play the direct or polite version of *Virtual Factory*, and then take a thirty-five-item test on the engineering concepts involved in the game (Wang et al. 2008). As shown in table 4.12, the main finding was that students who received the polite version of the game performed better on the learning test than did students who received the direct version, yielding an effect size of $d = 0.93$,

Table 4.12

Politeness principle: People learn better in games when an agent uses polite rather than direct wording

Source	Effect size
Wang et al. (2008)	0.93

which is a large effect. Overall, this study offers preliminary evidence for the *politeness principle in game design*: people learn better in games when on-screen agents use polite conversational style rather than direct conversational style. This principle is similar to the personalization principle found in *Design-a-Plant*. An important boundary condition is that this principle appears to be stronger for inexperienced learners than for experienced ones (Wang et al., 2008; Mayer, Johnson, Shaw, & Sandhu, 2006).

Which Features Improve Learning in **Cache 17*?***

What Is Cache 17*?*

Suppose you are on the trail of lost artwork that is hidden deep within a World War II bunker, and as you move through passageways you come to doors that are stuck closed. In order to open the doors, you must figure out how to create electric motors with the elements you find around you in the room, using a personal digital assistant (PDA) to find out the details of how the devices work (such as shown on the left side of figure 4.12). This is the approach taken in *Cache 17*, an adventure game developed by Koenig (2008). The instructional goal of the game is to help players understand how electric devices such as a wet battery cell work. The rationale for

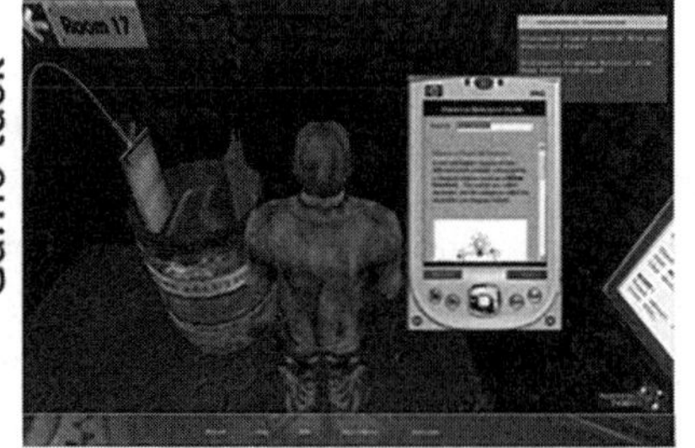

Figure 4.12

Do people learn better from playing *Cache 17* when there is a strong narrative theme?

embedding this scientific learning task within an adventure game is that the game's narrative theme may be motivating to learners, causing them to work harder and persist longer in understanding the material (Dickey, 2006).

Narrative Theme Principle: Do People Learn Better in Games with Strong Narrative Themes?

According to the narrative theme principle, players learn better from games with strong narrative themes. Figure 4.12 outlines the design of a value-added comparison aimed at testing this proposal. The base group receives a version of *Cache 17* without a strong narrative theme in which the player's goal is to move through the bunker system, but the lost artwork narrative is excluded. The enhanced group receives the same game, but also receives an introductory video describing the history of the lost artwork and asking the player to assume the role of an insurance claim agent. To assess the learning of how electric devices work, all players receive the same seventeen-item posttest containing both retention and transfer questions.

Table 4.13 shows that the results of a single experimental comparison carried out by Adams, Mayer, MacNamara, Koening, and Wainess (2012) in which adding a narrative theme to *Cache 17* resulted in slightly better posttest performance as compared to the base group. The effect size was $d = 0.16$, favoring the narrative group, which is a negligible effect. This result offers preliminary support for the *narrative theme principle for games*: people do not learn better in games with strong narrative themes. Given the limited amount of evidence, it is premature to rule out the instructional effectiveness of narrative theme in computer games.

Summary: What Works Based on the Value-Added Approach

In summarizing our work using the value-added approach, the top portion of table 4.14 lists the names and descriptions of seven promising principles

Table 4.13

Narrative theme principle: People do not learn better in games with strong narrative themes

Source	Effect size
Adams et al. (2012)	0.16

Table 4.14
Summary of value-added research: Which features improve learning?

Principle	Description	Game	Tests	Median effect size
Promising principles				
Modality	Use spoken words rather than printed words	*Design-a-Plant*	9	1.41
Personalization	Use conversational style rather than formal style	*Design-a-Plant*	5	1.58
Explanative feedback	Provide explanative feedback after key moves	*Circuit Game*	1	0.68
Self-explanation	Ask people to select an explanation for key moves	*Circuit Game*	3	0.90
Prompting	Prompt people to reflect on underlying rules	*Circuit Game*	2	0.65
Pretraining	Provide pretraining in key concepts	*Profile Game*	2	0.71
Politeness	Use polite wording rather than direct wording	*Virtual Factory*	1	0.93
Challenged principles				
Redundancy	Use redundant printed and spoken words rather than spoken words alone	*Design-a-Plant*	2	–0.22
Image	Put the agent's image on the screen	*Design-a-Plant*	4	0.22
Competition	Add competitive features to a game	*Circuit Game*	1	–0.06
Narrative theme	Provide a narrative theme for the game	*Cache 17*	1	0.16

of game design that have garnered some support in the examples of value-added research presented in this section: modality (d = 1.41 based on nine comparisons), personalization (d = 1.58 based on five comparisons), explanative feedback (d = 0.68 based on one comparison), self-explanation (d = 0.90 based on two comparisons), prompting (d = 0.65 based on two comparisons), pretraining (d = 0.71 based on two comparisons), and politeness (d = 0.93 based on one comparison).

The bottom portion of table 4.14 lists the names and descriptions of four challenged principles of game design that were not supported in the examples of value-added research presented in this section: redundancy

($d = -0.22$ based on two comparisons), image ($d = 0.22$ based on four comparisons), competition ($d = -0.06$ based on one comparison), and narrative theme ($d = 0.16$ based on one comparison).

These results should be viewed as preliminary because each principle is based on only one game tested by one lab group. Some are based on a single comparison, which clearly is not enough for drawing definitive conclusions. Chapter 5 provides a broader review of all available experimental comparisons from around the world that meet the methodological criteria for value-added game research as described in chapter 2.

Examples of Cognitive Consequences Research

Figure 4.13 summarizes the cognitive consequences approach to game research, which is intended to determine what is learned from playing an off-the-shelf computer game. Does playing a computer game improve your cognitive ability? Does it improve your knowledge and skill in academic domains? If these kinds of questions interest you, then you would benefit from game research that takes a cognitive consequences approach. As shown in figure 4.13, in game research based on the cognitive consequences approach, we compare the learning outcomes of people who play a commercially available game (game group) versus those who do not (control group). In cognitive consequences experiments, the main independent variable is whether or not people are required to play a game (or the amount

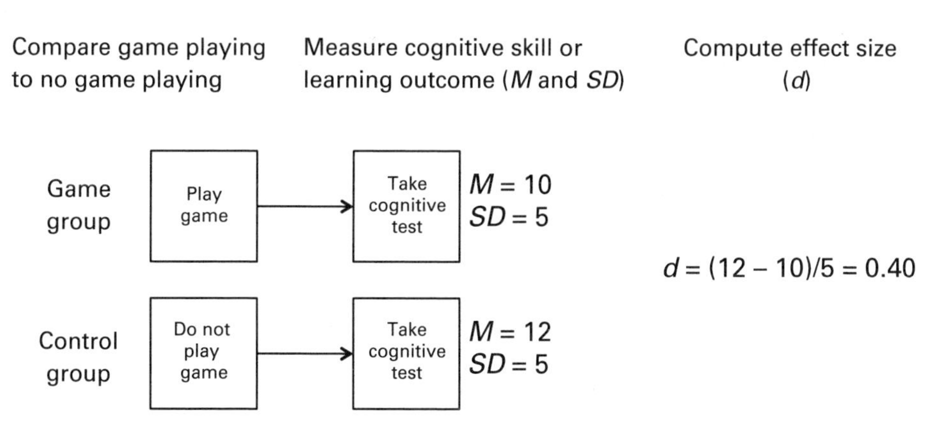

Figure 4.13
Cognitive consequences approach to game research

of time they are required to play it), and the main dependent measure assesses academic learning outcomes or cognitive skills necessary for academic success. The goal is to determine whether playing off-the-shelf games can have a positive cognitive impact on the game player. There is cause to take notice when the experimental comparisons result in an effect size greater than $d = 0.40$ on an important cognitive skill or learning outcome over the course of many experiments.

In this section, I provide examples of cognitive consequences research in two different venues: playing a variety of off-the-shelf games for an extended period within an after-school computer club and playing *Tetris* for an extended period in a research lab environment.

Positive Cognitive Consequences: What Do Students Learn from Playing Educational Games in an After-School Computer Club?

Figure 4.14 summarizes an experimental comparison between students who were assigned to play a variety of educational computer games within the context of an after-school computer club (game group) and equivalent students who were not initially invited to join the computer club (control

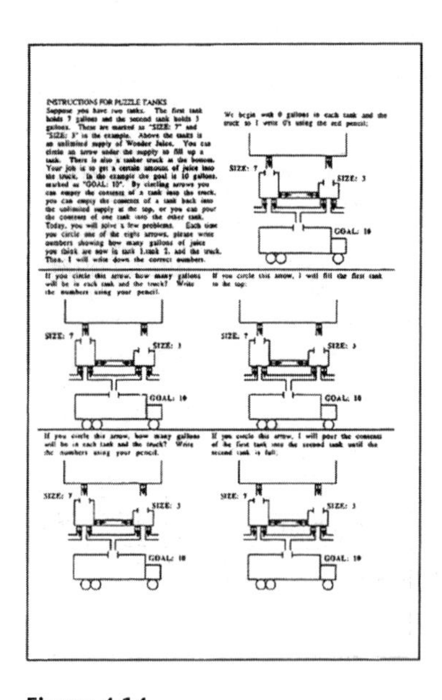

Game group: Play educational computer games in an after-school computer club for at least 10 weeks.

Control group: Matched students are randomly selected to not participate in an after-school computer club.

Transfer test: Solve math problems in a new educational game or translate word problems into equations.

Figure 4.14
What do students learn from playing educational games in an after-school computer club?

group). For the game group, we recruited a group of twenty-five fifth graders to participate in an after school computer club in which they learned how to follow directions to play a series of educational computer games in mathematics and reading. The club was called the Fifth Dimension, and was developed by Cole & Distributed Literacy Consortium (2006; Mayer et al., 1997). For the control group, we identified a group of twenty-five fifth graders who were matched for basic characteristics such as age, sex, teacher, English-language proficiency, and achievement test scores, but who were not initially invited to participate in an after-school computer club.

What are the cognitive consequences of regularly attending the computer club throughout the year and learning how to play a collection of educational games? To answer this question, we asked students in the game and control groups to learn to solve a new computer game—*Puzzle Tanks*—that involved developing efficient strategies for solving arithmetic problems, as shown on the left side of figure 4.14 (Mayer, Quilici, & Moreno, 1999). As summarized in the second row of table 4.15, the results indicated that the game group made fewer errors in solving problems in *Puzzle Tanks*, yielding an effect size of $d = 0.59$. In another study, shown in the first line table 4.15, the dependent measure was a math test that involved converting arithmetic word problems into equations (Mayer et al., 1997), yielding an effect size of $d = 0.87$ and favoring the game group. Overall, these are examples of cognitive consequences studies in which learning to play computer games resulted in learning skills that transferred to similar cognitive tasks. The two studies described in this section provide preliminary support for the idea that learning to play computer games can result in students learning academic skills that transfer to similar tasks.

Table 4.15

Positive cognitive consequences: People learn math skills from playing games in an after-school computer club

Source	Target skill	Effect size
Mayer et al. (1997)	Comprehend math problems	0.87
Mayer, Quilici, & Moreno (1999)	Solve math puzzles in a new game	0.59
Median		0.73

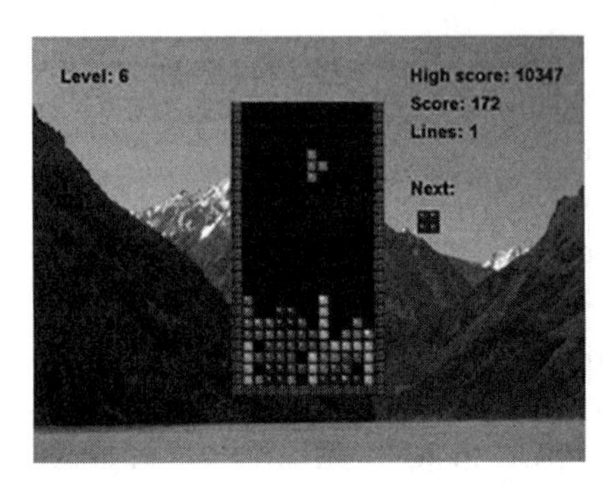

Game group: Novices play *Tetris* for 10 hours.

Control group: Matched students are randomly selected to not play *Tetris*.

Transfer test: Take a battery of spatial cognition tests.

Figure 4.15
What do students learn from playing *Tetris*?

Null Cognitive Consequences: What Do Students Learn from Playing* Tetris*?
Suppose the producers of a computer game claim that it will improve your cognitive skill. For instance, consider the claim that playing a spatial game like *Tetris*, which has been dubbed the greatest computer game of all time (McGonigal, 2011), will improve your spatial ability. As shown in figure 4.15, in playing *Tetris* you use a controller to rotate Tetris shapes in order to arrange them as they fall from the top of the screen.

Figure 4.15 summarizes a cognitive consequences study aimed at testing the claims for the cognitive consequences of playing *Tetris*. For the game group, we asked some college students who had never played *Tetris* to take a battery of spatial cognition tests as a pretest, play *Tetris* in our lab for ten one-hour sessions, and then take the same battery of spatial cognition tests as a posttest (Sims & Mayer, 2002). The spatial tests included computer-based measures of the speed of mental rotation of *Tetris* shapes, *Tetris*-like shapes, and letters as well as paper-based tests of spatial ability. For the control group, we identified a group of college students that had never played *Tetris* before and who were matched to the game-playing group on the basis of pretest scores. The control group took the same pretest and posttest as the game group, but did not play *Tetris* during the intervening time interval.

Students in both groups showed large and significant pretest-to-posttest improvements across each of nine tests of spatial ability. As highlighted in table 4.16, however, there was no strong evidence that students in the game group showed larger improvements than students in the control group, and

Table 4.16
Null cognitive consequences. People do not learn spatial cognition skills from playing *Tetris*

Source	Target skill	Effect size
Sims & Mayer (2002)	Mental rotation of non-*Tetris* letters	0.23
	Mental rotation of non-*Tetris* shapes	0.23
	Card rotation test	–0.15
Median		0.23

hence no evidence that ten hours of *Tetris* playing had useful cognitive consequences based on measures of spatial ability (Sims & Mayer, 2002). A more in-depth analysis showed that the game group tended to use a more efficient strategy for rotating *Tetris* shapes than did the control group. This research casts doubt on the idea that the specific skills learned in playing *Tetris* transfer broadly to other cognitive domains. In short, a major research question on using computer games to teach cognitive skills concerns the degree to which the game-based skills transfer to other cognitive domains.

Summary: What Works Based on the Cognitive Consequences Approach
Table 4.17 summarizes the main findings of our research concerning when playing off-the-shelf games improves learning outcomes. The top portion of the table shows that playing a collection of educational games over an extended period, such as within an after-school computer club, improved mathematics skills. The bottom portion shows that playing the puzzle game

Table 4.17
Summary of cognitive consequences research: When does game playing have positive consequences?

Description	Target skill	Tests	Median effect size
Promising venues			
Play a collection of educational games in an after-school computer club	Math	2	0.73
Unpromising venues			
Play a puzzle game (*Tetris*) in a lab environment	Mental rotation	3	0.23

Tetris for an extend period in a research lab did not improve spatial cognition skills such as mental rotation. This research should be viewed as examples of cognitive consequences research from a single lab rather than as definitive work. Chapter 6 provides a broader review of cognitive consequences game research from around the world that meets the criteria for scientific acceptability, as described in chapter 2.

Examples of Media Comparison Research

Figure 4.16 summarizes the design of media comparison experiments in game research. In media comparison studies, we compare the learning of academic content with one medium (such as with an interactive computer game) versus with another medium (such as a static slide presentation). The main independent variable is the medium that is used for delivering academic content to the learner (i.e., computer game versus nongame), and the main dependent measure taps the learning outcome. If games consistently produce effect sizes that average over $d = 0.40$, instructors might be inclined to consider using them as instructional tools.

It might be more accurate to consider games as an instructional format rather than a medium and reserve the term medium to refer to broader distinctions (such as between computer- based versus paper-based versus face-to-face contexts). For ease of communication, in this book I use the term *media* to refer to differences between instruction with computer games versus instruction with nongame multimedia presentations.

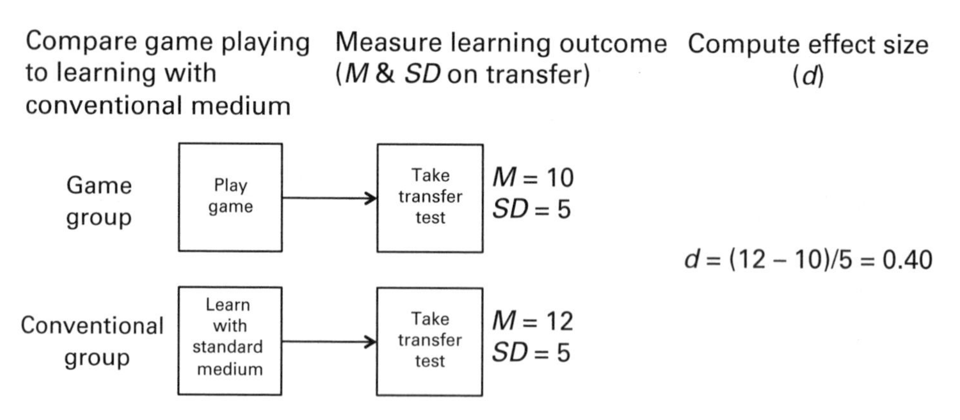

Figure 4.16
Media comparison approach to game research

The gamification principle is that students learn academic content better from games than from conventional media. A theoretical rationale for the gamification principle is that games prime intrinsic motivation in the players, causing them to initiate and maintain game-playing activity (Lepper & Malone, 1987; Malone & Lepper, 1987). In order to test this idea, in this section I explore whether people learn better from playing each of three games—*Design-a-Plant, Cache 17,* and *Crystal Island*—or from a computer-based presentation on the same material that does not involve games.

Positive Media Comparison: Do People Learn Better from Playing the Design-a-Plant *Game or a Conventional Lesson?*

Figure 4.17 summarizes the experimental comparison between the standard version of *Design-a-Plant* described in a previous section (game group) versus a computer-based multimedia lesson that contains the same graphics and words, but does not include an on-screen agent and does not allow the learner to design plants (conventional group). In the conventional group, the learner is shown the same problems as used in the game version, but then is shown the correct answers along with the same explanations as used in the game version. Then, the learner can click on a button to go on to the next segment.

Table 4.18 shows that across three different experiments, students who learned in the game version performed better on a transfer test than did students who learned with the conventional lesson, yielding a median effect size of $d = 0.97$, which is a large effect (Moreno et al., 2001). It appears that *Design-a-Plant* offers something that the conventional lesson does not

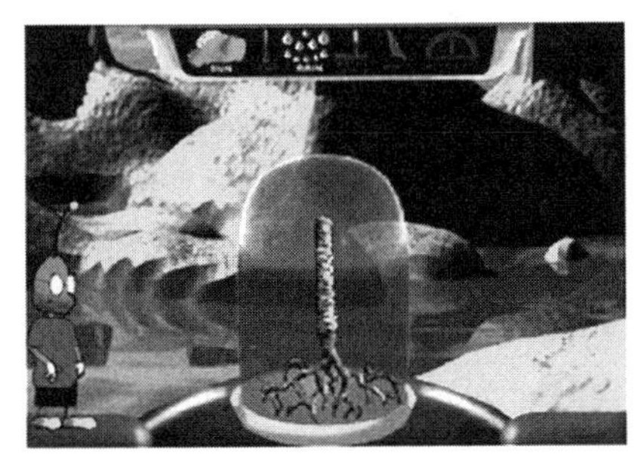

Game group: Choose roots, stem, and leaves. Interact with Herman-the-bug. Get narrated explanation.

Conventional group: Get narrated explanation.

Transfer test: Determine best environment for various plants.

Figure 4.17
Do people learn better from playing *Design-a-Plant* or an online lesson?

Table 4.18
Positive media comparison: People learn better from *Design-a-Plant* than from an online lesson

Source	Effect size
Moreno et al. (2001, expt. 1)	1.03
Moreno et al. (2001, expt. 2)	0.97
Moreno et al. (2001, expt. 3)	0.76
Median	0.97

offer, such as a more interesting and motivating learning context. This interpretation is supported by the finding that students in the game group rated the material as more interesting and expressed stronger preference to continue than did students in the conventional group. Overall, this finding provides preliminary evidence for the idea that game media can be more motivating, which I call the gamification principle, although better-controlled studies using the media comparison approach are needed.

Negative Media Comparison: Do People Learn Better from Playing Cache 17 *(or* Crystal Island*) or Watching a Slideshow?*

In spite of our initial success in finding support for the gamification principle, we were unable to confirm it in two additional well-controlled media comparison studies. First, figure 4.18 summarizes an experimental comparison between playing the standard version of *Cache 17* as described in a previous section (game group) versus receiving a slideshow that contains

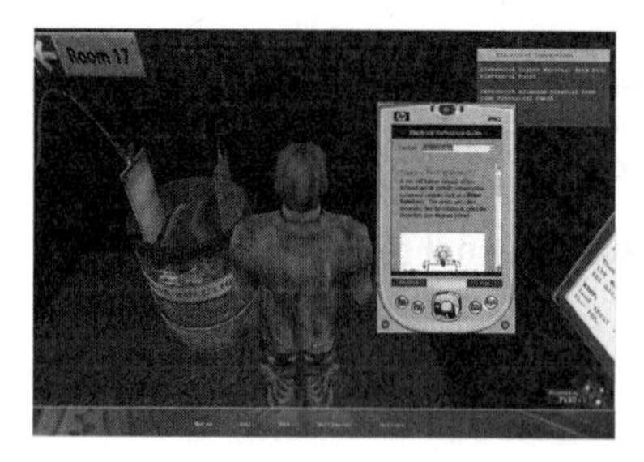

Game group: Solve a mystery problem by moving a character through a bunker system and building three electrical devices to open doors along the way.

Conventional group: Receive same information about electrical devices via a slideshow.

Transfer test: Write answers to questions about electric devices.

Figure 4.18
Do people learn better from playing *Cache 17* or watching a slideshow?

Table 4.19
Negative media comparison: People do not learn better from *Cache 17* than from a slideshow

Source	Effect size
Adams et al. (2012, expt. 2)	–0.57

the same target words and graphics pertaining to electric devices as in the game (conventional group). Although the slideshow presented the target information from the game, such as explanations from a PDA, it did not include any narrative about lost artwork or any requirement to navigate an avatar through a bunker system using a somewhat tedious interface. The theoretical rationale for the slideshow group is that deleting extraneous material reduces extraneous cognitive processing during learning, thereby freeing up cognitive capacity for essential and generative processing needed for academic learning.

Table 4.19 summarizes the results of Adams et al. (2012), in which the slideshow group performed much better than the control group on a transfer posttest ($d = -0.57$). Overall, the results do not support the gamification principle and are more consistent with the idea that extraneous game elements can compete for cognitive resources needed for meaningful learning.

Finally, figure 4.19 summarizes another experimental comparison between a game (game group) and slideshow (conventional group). The adventure game represented on the left side of figure 4.19 is *Crystal Island* (Spires, Turner, Rowe, Mott, & Lester, 2010), in which the participants must

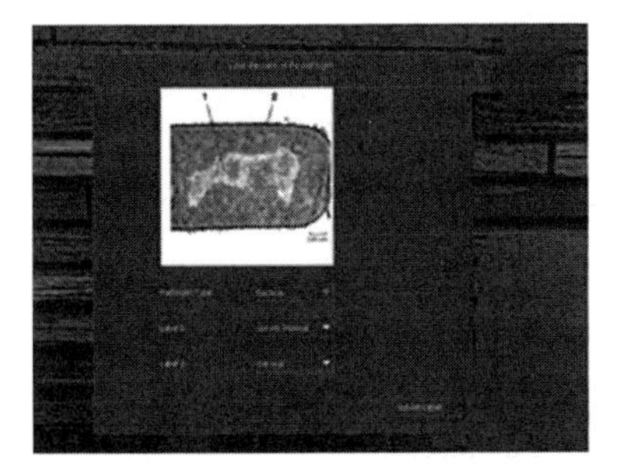

Game group: Figure out the cause of a disease on an island by gathering information and conducting tests with a microscope.

Conventional group: Receive same information about infectious diseases via a slideshow.

Transfer test: Write answers to questions about infectious diseases.

Figure 4.19
Do people learn better from playing *Crystal Island* or watching a slideshow?

Table 4.20
Negative media comparison: People do not learn better from *Crystal Island* than from a slideshow

Source	Effect size
Adams et al. (2012, expt. 1)	–0.31

determine the cause of a disease outbreak on a remote island by communicating with in-game characters, using lab microscopes to collect data, and reading in-game charts placed in various rooms. All communication was via printed on-screen text, and the game included three quizzes on the target material given at various points in the game. In contrast, the slideshow group received all the same words and images about infectious disease from the game, and took the same quizzes as in the game, but did not receive any extraneous narrative about *Crystal Island* or images of characters. Table 4.20 shows that Adams et al. (2012) found that the conventional group performed better on a transfer posttest than did the game group (d = –0.31), thereby offering no support for the gamification principle.

Summary of Media Comparison Research: When Are Games Better Than Conventional Media?

Table 4.21 summarizes the results of our media comparison studies in which *Design-a-Plant* was more effective than conventional media (d = 0.97

Table 4.21
Summary of media comparison research: When are games more effective than conventional media?

Game	Conventional group	Content	Tests	Median effect size
Games are better				
Design-a-Plant	Computer-based presentation	Botany	3	0.97
Conventional media are better				
Cache 17	Computer-based slideshow	Electric devices	1	–0.57
Crystal Island	Computer-based slideshow	Infectious disease	1	–0.31

based on three comparisons), but conventional media were more effective than *Cache 17* (*d* = –0.57 based on one comparison) and *Crystal Island* (*d* = –0.31 based on one comparison). These results should be interpreted as examples of media comparison research rather than as offering definitive conclusions. A broader review of available experiments that appropriately use the media comparison approach is presented in chapter 7.

As noted in chapter 2, Clark (2001) argues that conducting media comparison studies may not be a fruitful research approach for most aspects of instructional technology. The rationale is that it is not the instructional medium that causes learning but rather it is the instructional method that causes learning. In media comparison research, it is easy to confound medium and method, thereby making it impossible to attribute learning effects to games per se. Media comparison research should be viewed with an eye toward Clark's admonitions.

Advantages and Disadvantages of Each Genre of Game Research

Which research approach should be used in game research—value added, cognitive consequences, or media comparison? It is widely recognized that the choice of research methodology should depend on the research question (Shavelson & Towne, 2002), so there is a place for each approach.

First, when you want to know whether an instructional feature should be added to a game, as I often do, the value-added approach is appropriate. In short, the value-added approach should be used to help answer instructional design questions about which features help people to learn from educational games. On the practical side, the value-added research has important practical implications for the science of instruction—by contributing research-based principles of game design. It is relevant to educational game designers who wish to decide how to design effective educational games. On the theoretical side, value-added research has important theoretical implications for the science of learning—by testing whether theory-based features have the predicted effects on game design. Thus, the value-added approach is a useful tool for learning theorists who wish to explain how learning works.

Second, when you want to know whether playing a particular off-the-shelf game has desirable effects on learners, the cognitive consequences approach is appropriate. On the practical side, the cognitive consequences approach has important implications for the science of instruction—by contributing to recommendations for how to use games to promote the learning of cognitive skills relevant to academic learning. On the theoretical side, the cognitive consequences approach has important theoretical implications for the science of learning—particularly concerning the nature of transfer from game skills to real-world skills.

Third, when you want to determine whether games are more effective learning environments than conventional media, media comparison research is appropriate. You would be taking a media comparison approach in spite of warnings about the difficulty in separating the learning effects of instructional media from instructional methods and in equating the two groups on all dimensions except gaming. On the practical side, given cost considerations, it would be useful to determine whether students learn as effectively from a text-based lesson as from a computer-based game. In addition, media comparison research could help determine whether people can learn something of academic value in their spare time by playing games in an informal learning environment even if games are not as effective as conventional media used in formal learning environments. On the theoretical side, media comparison research could help determine whether different media can be seen as equivalent delivery systems or whether games afford motivational properties not found in more formal environments.

In short, there is a place for all three kinds of research genres, although the value-added approach is most appropriate to answer instructional design questions. Informal, observational studies can be added to help provide richness and context.

Note

Portions of this chapter are based on Mayer (2011).

References

Note: Asterisk (*) indicates study is in review database.

Abt, C. C. (1970). *Serious games*. New York: Viking.

*Adams, D. M., Mayer, R. E., MacNamara, A., Koening, A., & Wainess, R. (2012). Narrative games for learning: Testing the discovery and narrative hypothesis. *Journal of Educational Psychology, 104*, 235–249.

Brown, P., & Levinson, S. C. (1987). *Politeness*. New York: Cambridge University Press.

Clark, L., Lawrence, A. J., Astley-Jones, F., & Gray, N. (2009). Gambling near-misses enhance motivation to gamble and recruit win-related brain circuitry. *Neuron, 61*(3), 481–490.

Clark, R. E. (2001). *Learning from media*. Greenwich, CT: Information Age Publishing.

Cole, M., & Distributed Literacy Consortium. (2006). *The fifth dimension*. New York: Russell Sage Foundation.

Connolly, T. M., Boyle, E. A., MacArthur, E., Hainey, T., & Boyle, J. M. (2012). A systematic review of empirical evidence on computer games and serious games. *Computers & Education, 59*, 661–686.

*DeLeeuw, K., & Mayer, R. E. (2011). Cognitive consequences of making computer-based learning activities more game-like. *Computers in Human Behavior, 27*, 2011–2016.

Dessouky, M. M., Verma, S., Bailey, D., & Rickel, J. (2001). A methodology for developing a Web-based factory simulator for manufacturing education. *IEEE Transactions, 33*, 167–180.

Dickey, M. D. (2006). Game design narrative for learning: Appropriating adventure game design narrative devices and techniques for the design of interactive learning environments. *Educational Technology Research and Development, 54*(3), 245–263.

*Fiorella, L., & Mayer, R. E. (2012). Paper-based aids for learning with a computer-based game. *Journal of Educational Psychology*, 104, 1074–1082.

Fonseca, B. A., & Chi, M.T.H. (2011). Instruction based on self-explanation. In R. E. Mayer & P. A. Alexander (Eds.), *Handbook of research on learning and instruction* (pp. 296–321). New York: Routledge.

Hattie, J. (2009). *Visible learning*. New York: Routledge.

Hattie, J., & Gan, M. (2011). Instruction based on feedback. In R. E. Mayer & P. A. Alexander (Eds.), *Handbook of research on learning and instruction* (pp. 249–271). New York: Routledge.

Hayes, R. T. (2005). *The effectiveness of instructional games: A literature review and discussion*. Naval Air Warfare Center Training Systems Division, Technical Report 2005–004, Orlando, FL.

Honey, M., & Hilton, M. (Eds.). (2011). *Learning science through computer games and simulations*. Washington, DC: National Academy Press.

*Johnson, C. I., & Mayer, R. E. (2010). Adding the self-explanation principle to multimedia learning in a computer-based game-like environment. *Computers in Human Behavior, 26*, 1246–1252.

Koenig, A. D. (2008). *Exploring effective educational video game design: The interplay between narrative and game-schema construction.* (Unpublished doctoral dissertation). Arizona State University, Department of Psychology, Phoenix.

Lepper, M. R., & Malone, T. W. (1987). Intrinsic motivation and instructional effectiveness in computer-based education. In R. E. Snow & M. J. Farr (Eds.), *Aptitude, learning, and instruction, Volume 3: Conative and affective process analyses* (pp. 255–286). Hillsdale, NJ: Erlbaum.

Lester, J. C., Stone, B. A., & Stelling, G. D. (1999). Lifelike pedagogical agents for mixed-initiative problem solving in constructivist learning environments. *User Modeling and User-Adapted Interaction, 9*, 1–44.

Malone, T. W. (1981). Toward a theory of intrinsically motivating instruction. *Cognitive Science, 4*, 333–369.

Malone, T. W., & Lepper, M. R. (1987). Making learning fun: A taxonomy of intrinsic motivation for learning. In R. E. Snow & M. J. Farr (Eds.), *Aptitude, learning, and instruction, Volume 3: Conative and affective process analyses* (pp. 223–253). Hillsdale, NJ: Erlbaum.

Mayer, R. E. (2009). *Multimedia learning* (2nd ed.). New York: Cambridge University Press.

Mayer, R. E. (2011). Multimedia learning and games. In S. Tobias & J. D. Fletcher (Eds.), *Computer games and instruction* (pp. 281–305). Greenwich, CT: Information Age Publishing.

Mayer, R. E., & DaPra, S. (2012). An embodiment effect in computer-based learning with animated pedagogical agents. *Journal of Experimental Psychology: Applied, 18*, 239–252.

*Mayer, R. E., & Johnson, C. I. (2010). Adding instructional features that promote learning in a game-like environment. *Journal of Educational Computing Research, 42*, 241–265.

Mayer, R. E., Johnson, W. L., Shaw, E., & Sandhu, S. (2006). Constructing computer-based tutors that are socially sensitive: Politeness in educational software. *International Journal of Human-Computer Studies, 64*, 36–42.

*Mayer, R. E., Mautone, P. D., & Prothero, W. (2002). Pictorial aids for learning by doing in a multimedia geology simulation game. *Journal of Educational Psychology, 94*, 171–185.

*Mayer, R. E., Quilici, J. H., & Moreno, R. (1999). What is learned in an after-school computer club? *Journal of Educational Computing Research, 18,* 223–235.

*Mayer, R. E., Quilici, J. H., Moreno, R., Duran, R., Woodbridge, S., Simon, R., et al. (1997). Cognitive consequences of participation in a Fifth Dimension after-school computer club. *Journal of Educational Computing Research, 16,* 353–370.

McGonigal, J. (2011). *Reality is broken: Why games make us better and how they can change the world.* New York: Penguin Press.

*Moreno, R., & Mayer, R. E. (2000). Engaging students in active learning: The case for personalized multimedia messages. *Journal of Educational Psychology, 93,* 724–733.

*Moreno, R., & Mayer, R. E. (2002a). Learning science in virtual reality environments: Role of methods and media. *Journal of Educational Psychology, 94,* 598–610.

*Moreno, R., & Mayer, R. E. (2002b). Verbal redundancy in multimedia learning: When reading helps listening. *Journal of Educational Psychology, 94,* 156–163.

*Moreno, R., & Mayer, R. E. (2004). Personalized messages that promote science learning in virtual environments. *Journal of Educational Psychology, 96,* 165–173.

*Moreno, R., Mayer, R. E., Spires, H. A., & Lester, J. (2001). The case for social agency in computer-based teaching: Do students learn more deeply when they interact with animated pedagogical agents? *Cognition and Instruction, 19,* 177–213.

Nass, C., & Brave, S. (2005). *Wired for speech.* Cambridge, MA: MIT Press.

O'Neil, H. F., Wainess, R., & Baker, E. L. (2005). Classification of learning outcomes: Evidence from the computer games literature. *Curriculum Journal, 16,* 455–474.

Perez, R. S. (2008). Summary and discussion. In H. F. O'Neil & R. S. Perez (Eds.), *Computer games and team and individual learning* (pp. 287–306). Amsterdam: Elsevier.

Prothero, W. (1997). *Our dynamic planet.* [Computer program.] Santa Barbara: Department of Geological Sciences, University of California at Santa Barbara.

Randel, J. M., Morris, B. A., Wetzel, C. D., & Whitehill, B. V. (1992). The effectiveness of games for educational purposes: A review of recent research. *Simulation and Gaming, 23,* 261–276.

Reeves, B., & Nass, C. (1996). *The media equation.* New York: Cambridge University Press.

Rieber, L. P. (2005). Multimedia learning in games, simulations, and microworlds. In R. E. Mayer (Ed.), *The Cambridge handbook of multimedia learning* (pp. 549–568). New York: Cambridge University Press.

Roy, M., & Chi, M.T.H. (2005). The self-explanation principle in multimedia learning. In R. E. Mayer (Ed.), *The Cambridge handbook of multimedia learning* (pp. 271–286). New York: Cambridge University Press.

Shavelson, R. J., & Towne, L. (Eds.). (2002). *Scientific research in education.* Washington, DC: National Academy Press.

Shute, V. J. (2008). Focus on formative feedback. *Review of Educational Research, 78,* 153–189.

*Sims, V. K., & Mayer, R. E. (2002). Domain specificity of spatial expertise: The case of video game players. *Applied Cognitive Psychology, 16,* 97–115.

Sitzmann, T. (2011). A meta-analytic examination of the instructional effectiveness of computer-based simulation games. *Personnel Psychology, 64,* 489–528.

Spires, H. A., Turner, K. A., Rowe, J., Mott, B., & Lester, J. (2010, May). *Game-based literacies and learning: Towards a transactional theoretical perspective.* Paper presented at the annual meeting of the American Educational Research Association, Denver, CO.

Tobias, S., Fletcher, J. D., Dai, D. Y., & Wind, A. P. (2011). Review of research on computer games. In S. Tobias & J. D. Fletcher (Eds.), *Computer games and instruction* (pp. 127–222. Charlotte, NC: Information Age Publishing.

Vogel, J. J., Vogel, D. S., Cannon-Bowers, J., Bowers, C. A., Muse, K., & Wright, M. (2006). Computer gaming and interactive simulations for learning: A meta-analysis. *Journal of Educational Computing Research, 34,* 229–243.

*Wang, N., Johnson, W. L., Mayer, R. E., Rizzo, P., Shaw, E., & Collins, H. (2008). The politeness effect: Pedagogical agents and learning outcomes. *International Journal of Human-Computer Studies, 66,* 96–112.

Young, M. F., Slota, S., Cutter, A. B., Jalette, G., Mullin, G., Lai, B., et al. (2012). Our princess is in another castle: A review of trends in serious gaming for education. *Review of Educational Research, 82,* 61–89.

5 Value-Added Approach: Which Features Improve a Game's Effectiveness?

Chapter Outline

Introduction

- Rationale for the Value-Added Approach to Game Research
- Rationale for the Evidence-Based Approach to Game Research
- Theoretical Framework

Method

- Evidence Collection
- Evidence Selection
- Evidence Coding
- Evidence Summarization and Interpretation

Results

- What Works: Five Promising Features
- What Does Not Work: Two Unpromising Features
- What Has Not Yet Been Shown to Work: Six Not-Yet-Promising Features

Discussion

- Practical Contributions
- Theoretical Contributions
- Methodological Contributions
- Future Contributions

Summary

The value-added approach to game research compares the learning outcome performance of students who learned by playing a game versus students who were assigned to play the same game with one instructional feature added. A review of value-added game research identified five promising features that tend to improve learning from computer games: modality (d = 1.41 based on nine comparisons), personalization (d = 1.54 based on eight comparisons), pretraining (d = 0.75 based on seven comparisons), coaching (d = 0.68 based on seven comparisons), and self-explanation (d = 0.81 based on six comparisons). Two features were unpromising: immersion (d = –0.14 based on six comparisons) and redundancy (d = –0.23 based on two comparisons). Six features have not yet been shown to be promising based on one to four comparisons: competition (d = –0.16), segmenting (d = 0.23), image (d = 0.22), choice (d = 0.30), narrative theme (d = 0.38), and learner control (d = 0.19). Overall, the value-added approach is a useful methodology for identifying effective features of computer games for learning.

Introduction

Suppose you were asked to design a computer game to help students achieve an educational goal, such as learning the principles of electric circuits or how to solve math problems? Although you would want to include entertaining features that make the game enjoyable, how would you decide which instructional features to include to make the game educationally effective? In short, how should educational computer games be designed to promote learning? The goal of this review is to summarize research evidence that pinpoints which instructional features improve learning when they are added to an educational game or simulation.

Games for learning are computer games and simulations that are intended to promote learning in academic areas (e.g., science, mathematics, engineering, etc.). For the purposes of this review, educational games or games for learning refers to both games and simulations. The primary goal of this review is to examine and summarize what the research evidence has to say about how to design educational games to promote learning. In particular, this review focuses on one basic genre of game research—value-added research, which compares the learning outcomes of students assigned to play a base version of a game versus the same game with an added instructional feature (Mayer, 2011). The field of game effectiveness research is in its early phase, so this review seeks to provide needed guidance in terms of promising research directions and methods as well as practical implications for the design of effective games.

Research on educational games reflects a tension between entertaining features of games, which promote motivation and create fun experiences, versus instructional features of games, which promote appropriate cognitive processing during learning and lead to meaningful learning outcomes (Rittenfeld & Weber, 2006). The promise of games for learning rests in their motivational appeal, which instigates and maintains the learner's game playing. The challenge for game designers is to infuse enough instructional elements to guide the learner's cognitive processing during game play toward the construction of meaningful learning outcomes, while not completely destroying the fun of the game. This review is intended to help meet this challenge to identify effective instructional features of educational games.

Rationale for the Value-Added Approach to Game Research

The value-added approach to game research involves comparing the learning outcome score of a group that was assigned to play a base version of a game as opposed to the learning outcome score of a group that was assigned to play the same game with one instructional feature added. The value-added approach is designed to answer the instructional design question, "Does adding feature X improve learning?" The rationale for the value-added approach is that making controlled experimental comparisons is widely recognized as the most efficient way to test for instructional effectiveness in educational research (Phye, Robinson, & Levin, 2005; Shavelson & Towne, 2002).

The present review concentrates solely on studies using the value-added approach because it is the most appropriate methodology for answering instructional design questions about which game features promote learning. In contrast, observational studies seek to describe game play or game players, cognitive consequences studies seek to determine whether game playing affects cognitive skills, and media comparison studies seek to determine whether game playing is an effective instructional venue, as described in chapter 1.

Previous comprehensive reviews of game research have mixed together many different genres of game research including value-added, observational, cognitive consequences, and media studies in order to provide an overview of the field (Connolly, Boyle, MacArthur, Hainey, & Boyle, 2012; Hannafin & Vermillion, 2008; Hayes, 2005; Honey & Hilton, 2011; Randel, Morris, Wetzel, & Whitehill, 1992; Tobias, Fletcher, Dai, & Wind, 2011; Vogel et al., 2006; Young et al., 2012). The present review differs from previous ones by focusing only on value-added studies in order to better detect game features that have the potential to improve learning. This more focused approach is intended to identify promising game features that have been shown to improve learning from games and unpromising game features that have been shown to not improve learning games.

In particular, this review builds on previous work in the field of game research by: (1) focusing specifically on the most appropriate genre of game research, thereby reviewing studies that use a well-specified experimental design; (2) applying specific criteria for selecting clear-cut experimental comparisons based on learning outcome measures in academic areas, thereby keeping the spotlight on instructional effectiveness; (3) examining boundary conditions for game effectiveness suggested by cognitive theories of multimedia learning, thereby grounding the investigation in cognitive theory; and (4) including up-to-date studies on the educational effectiveness of computer games, thereby keeping up with the pace of game research.

Rationale for the Evidence-Based Approach to Game Research

Proponents make strong claims for using computer games to revolutionize education, but it is worthwhile to determine whether such claims are supported by research evidence. Specifically, evidence is needed to guide game designers who wish to create games for learning. O'Neil and Perez (2008) eloquently make this point in *Computer Games and Team and Individual Learning*: "While effectiveness of game environments can be documented in terms of intensity and longevity of engagement … as well as the commercial success of the games, there is much less solid empirical information about what outcomes are systematically achieved by the use of individual and multiplayer games to train participants in acquiring knowledge and skills. Further, there is almost no guidance for game designers and developers on how to design games that facilitate learning" (p. ix). This review is intended to help provide that needed guidance.

Visionaries foresee a future in which computer games revolutionize education. In *Don't Bother Me Mom—I'm Learning*, Prensky (2006) suggests that "kids learn more positive, useful things for their future from their video games than they learn in school" (p. 4). As Gee (2007) maintains in *Good Video Games and Good Learning*, "Good games are problem-solving spaces that create deep learning, learning that is better than what we often see today in our schools" (p. 10). In *How Computer Games Help Children Learn*, Shaffer (2006) observes that "the key to solving the current crisis in education will be to use the power of computer and video games to give all children access to experiences and build interest and understanding" (p. 8). Similarly, in *Reality Is Broken*, McGonigal (2011) states, "I foresee games that fix our educational systems" (p. 14).

In a recent game review based on over three hundred articles, Young et al. (2012) found "little support for the academic value of video games in science and math" (p. 61). Yet in a rebuttal, Tobias and Fletcher (2012) chastised the authors for ignoring relevant research: "Young et al. reviewed research relating games to achievement in science but did not review a program of research by Mayer and colleagues [summarized in Mayer, 2011]. That research reported data from a number of studies that found near transfer effects in games dealing with principles underlying plant growth or electrical circuits" (p. 234). As summarized in chapter 4, in that preliminary line of research, the value-added approach has been especially promising in helping pinpoint game features that promote learning.

Overall, there appears to be a huge gap between claims for the educational effectiveness of games and the conclusions of scientific literature reviews. The goal of this review is to help bridge this gap by conducting fine-grained and focused meta-analyses on game research based on the value-added approach—carefully examining each experimental comparison that meets our rigorous criteria.

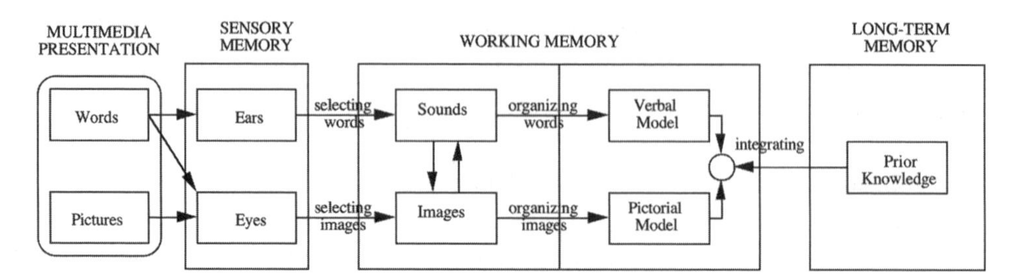

Figure 5.1
Cognitive theory of multimedia learning.

Theoretical Framework

As detailed in chapter 3, a cognitive theory of learning is helpful in understanding which game features might improve learning, under which conditions, and for which learners. The present review starts with the cognitive theory of multimedia learning (Mayer, 2005, 2009) and related ideas from cognitive load theory (Sweller, 2005; Sweller, Ayres, & Kalyuga, 2011), which have been useful in explaining principles for the design of multimedia instruction. According to the cognitive theory of multimedia learning, as shown in figure 5.1, information from the outside world enters the learner's cognitive system through the eyes and ears, and is held briefly in a sensory memory. If the learner attends to the incoming information (indicated by the "selecting" arrows), it is transferred to working memory for further processing. In working memory, the learner may mentally organize the material into coherent representations (indicated by the "organizing" arrows), and mentally connect the representations with each other as well as with relevant prior knowledge activated from long-term memory (indicated by the "integrating" arrows). Three important aspects of the cognitive theory of multimedia learning are: dual channels—people have separate channels for processing visual and verbal material; limited capacity—people can engage in only a small amount of cognitive processing in each channel at any one time; and active processing—meaningful learning depends on the learner engaging in appropriate cognitive processing during learning, including selecting, organizing, and integrating.

During learning from games, there are three major demands on the learner's cognitive system:

- *Extraneous processing* is cognitive processing that does not serve the instructional goal and is caused by poor instructional design. Games with attention-grabbing graphics, dazzling special effects, and intense realism can instigate extraneous processing, as can complicated interfaces or continual motor activity.
- *Essential processing* is cognitive processing intended to mentally represent the essential material and is caused by the complexity of the essential material. For example, essential processing is needed in games aimed at teaching about systems with many interacting parts, such as Ohm's law in electricity or Boyle's law in chemistry.
- *Generative processing* is cognitive processing aimed at deeper understanding and is caused by the learner's motivation. For instance, generative processing is needed in simulation games in which the learner takes on the role of a farmer who must make decisions.

An important instructional implication is that learners may face three kinds of instructional scenarios:

- *Extraneous overload* occurs when the amount of extraneous processing is so great that the learner is not able to engage in enough essential and generative processing. The instructional solution is to reduce extraneous processing. Reducing the level of physical realism, for example, can help the learner focus on the essential academic content.
- *Essential overload* occurs when the amount of essential processing is so great that the learner is not able to engage in enough of it. The instructional solution is to manage essential processing. For example, providing pretraining about the key elements in a game can help the learner concentrate on the relations among the elements.
- *Generative underutilization* occurs when capacity is available for generative processing, but the learner does not engage in it. The instructional solution is to foster generative processing. Using a conversational style can help motivate learners to work harder to make sense of the rules underlying a game.

For each instructional feature examined in this review, it is useful to assess the degree to which it addresses the instructional goal of reducing extraneous processing, managing essential processing, and fostering generative processing. This theoretical grounding can be useful in predicting boundary conditions for the effectiveness of each instructional feature. The theoretical basis for game research is discussed more fully in chapter 3.

Method

The primary goal of this value-added review is to survey and summarize the available evidence on which instructional features are most promising for improving academic learning with games along with the conditions under which these instructional features are most effective. In this review, I analyze all available published papers in which learning outcomes are compared between a group that plays a base version of an educational game (control group) and a group that plays the same game with one added instructional feature (experimental group). The procedure for this value-added review includes evidence collection, evidence selection, evidence coding, and evidence summarizing.

Evidence Collection

During evidence collection, I conducted searches of major social science online databases such as PsychINFO and ERIC, using appropriate keywords such as "computer games," "video games," "serious games," "educational games," "simulation games," and "online games." I also searched the listed references of previous literature reviews (e.g., Connolly et al., 2012; Honey & Hilton, 2011; Randel et al., 1992; Tobias et al., 2011; Vogel et al., 2006; Young et al., 2012), searched the listed references of recent relevant research papers; and conducted a search of all "cited by" papers for classic relevant papers and reviews.

Overviews of the research literature on computer games tend to show that the vast majority of the available papers on computer games do not report scientifically rigorous research evidence (Clark, Yates, Early, & Moulton, 2011). For instance, although O'Neil, Wainess, and Baker (2005) identified four thousand articles on computer games published in peer-reviewed journals, only nineteen were found that reported original data based on scientifically valid studies. On the other hand, a recent review by Tobias, Fletcher, Dai, and Wind (2011) that used somewhat-broader criteria and included more recent work was able to identify nearly a hundred studies contributing to the evidence base on learning with computer games, whereas a recent review with even broader criteria netted over three hundred studies (Young et al., 2012). In the present review, I found that the field has been able to produce a substantial corpus of game research papers using the valued-added approach that meet rigorous scientific criteria.

Evidence Selection

During evidence selection, I eliminated all papers that did not meet the following criteria:

- The independent variable involves a comparison of a control group that is assigned to play a base version of a game versus a treatment group that is assigned to play the same game with one additional instructional feature added. This can be seen as addressing the criteria of experimental control and random assignment, as described in chapter 2.
- The dependent measure involves a measure of learning outcome performance such as answering questions or solving problems (rather than self-reports or in-game activity). This can be seen as the criterion of appropriate measurement, as discussed in chapter 2.
- The paper reports the mean (M), standard deviation (SD), and sample size (n) for a measure of learning outcome performance for each group, or contains other statistical information that allows for computing a value of effect size (e.g., d or r). This also can be seen as an aspect of the criterion of appropriate measurement, as explained in chapter 2.
- The instructional content is in an academic area.
- The paper is published in a peer-reviewed research journal or book that is accessible.

Forays into the game literature suggest that the vast majority of papers on educational games do not report original research evidence. I recognize that experts tend to call for including papers from conferences, technical reports, dissertations, and other nonpublished sources in order to avoid the file-drawer problem in which papers with large and significant differences may have a better chance of getting published (Ellis, 2010; Lipsey & Wilson, 2001; Rosenthal, Rosnow, & Rubin, 2000). Given the particular problem of low quality in educational research (Mosteller & Boruch, 2002; Phye et al., 2005), however, I opted to focus on papers that meet the five criteria of research quality and appropriateness.

Evidence Coding

During evidence coding, I created a database for each experimental comparison between a group that receives a base version of game versus a group that receives the same game with one instructional feature added. The coding included descriptive cells such as a full publication citation (with experiment number), description of the game and game-playing environment, description of the added feature, description of the test, and description of subjects; categorical cells such as type of added feature (e.g., coaching, pretraining, personalization, immersion, choice, learner control, self-explanation prompts, narrative theme, competition, segmenting, modality, etc.), type of dependent measure (retention, transfer, etc.), type of game (e.g., action, puzzle, simulation, etc.), type of content (e.g., science, mathematics, history, etc.), age category (pre-K, elementary, secondary, college, adult), and type of setting (classroom, lab, home, etc.); and numerical cells (e.g., mean score, standard deviation, and sample size of each group; and effect size based on Cohen's *d*).

When there were multiple measures of retention and/or multiple measures of transfer, I recorded them all in the database and used the most representative measure of each. An alternative is to use the average of all retention measures and the average of all transfer measures (Ellis, 2010). In the interests of simplicity and in concert with previous meta-analyses of instructional effectiveness (Mayer, 2009), I opted to focus primarily on one measure of transfer (if available) for each comparison, because transfer is widely recognized as a paramount educational goal (Anderson et al., 2001).

In addition, a separate entry was made for each major subgroup (such as males versus females, high prior knowledge versus low prior knowledge, ethnic groups, etc.) or game-playing venue in order to identify boundary conditions that determine when or for whom a particular instructional feature has its strongest effects. It is possible to have more than one experiment in a paper, so each distinct experimental comparison had its own entry in the database as long as it was based on different data.

Evidence Summarization and Interpretation

During evidence summarization, I computed effect size for each experimental comparison using Cohen's (1988) *d*. I recognize that some experts recommend computing effect size based on *r* or other measures in the *r* family, such as omega squared or eta squared (Rosenthal et al., 2000) on the grounds that correlation-based measures are more general. I prefer to use *d* because all comparisons in this review are between two qualitatively different groups, *d* is somewhat more intuitive to communicate, and *d* can easily be converted into *r*. Where it is not possible to compute *d*, I computed Glass's delta (which is based on using the standard deviation of the control group as the denominator). I recognize that Hedges's *g* is sometimes recommended when there are large differences in sample size (Ellis, 2010), but I opted to forego this adjustment in the interests of simplicity and consistency. Finally, some experts call for adjusting dependent measures based on test reliability (Ellis, 2010), but this is not feasible because most studies do not include measures of test reliability.

For each instructional feature, I computed the median effect size. I recognize that experts in meta-analysis call for using weighted measures of *d* based on sample size in computing a mean effect size across multiple studies (Ellis, 2010), but I opted to report median effect size in the interests of simplicity and to avoid overweighing any one particular study. I focused on transfer test scores for computing median effect size, in light of its importance for assessing educational effectiveness.

Results

Table 5.1 summarizes the effectiveness of each instructional feature for which there is an evidence base by providing the name of the feature, description, number of experimental comparisons, and median effect size based on a measure of transfer.

What Works: Five Promising Features

The primary goal of this review is to identify promising features that have been shown to improve learning when added to a game. The top section of table 5.1 lists five promising features, each of which improves transfer performance by at least 0.5 standard deviations (i.e., yields a median effect size greater that $d = 0.5$) based on at least six experimental comparisons: modality, personalization, pretraining, coaching, and self-explanation.

Modality Principle

The first row in table 5.1 lists the modality principle for games: people learn better from a game when the words are spoken rather than printed.

Table 5.1
Research using the value-added approach

Name	Brief description	Comparisons	Median effect size
Promising features			
Modality	Present words in spoken form	9 of 9	1.41
Personalization	Use conversational style	8 of 8	1.54
Pretraining	Provide pregame experiences	7 of 7	0.77
Coaching	Provide advice or explanations	6 of 7	0.68
Self-explanation	Provide prompts to explain	5 of 6	0.81
Not-yet-promising features			
Competition	Show score for competition	1 of 3	–0.06
Segmenting	Break screen into parts	2 of 4	0.23
Image	Include agent's image on-screen	3 of 4	0.22
Narrative theme	Add engaging story line	2 of 2	0.38
Choice	Allow learner to choose format	3 of 3	0.30
Learner control	Allow learner to control order	0 of 1	0.19
Unpromising features			
Immersion	Use virtual reality	3 of 6	–0.14
Redundancy	Provide printed and spoken words	0 of 2	–0.23

The theoretical rationale for the modality principle is that the learner's visual channel can become overloaded in playing a game that includes printed text on the screen. When words are presented in spoken form, some of the processing demand can be off-loaded from the visual channel to the verbal channel, which has processing capacity available. In this way, the learner can effectively increase processing capacity by making better use of the verbal channel. Thus, the modality principle represents an example of managing essential processing in the cognitive theory of multimedia learning, as described in chapter 3.

The modality principle was upheld in nine out of nine experimental comparisons, yielding a median effect size of $d = 1.41$, which is a large effect. Overall, this review shows that the modality principle, which has been established as a major principle for the design of multimedia lessons (Ginns, 2005; Low & Sweller, 2005; Mayer, 2009), also applies to the design of educational games.

Table 5.2 summarizes each of nine experimental comparisons that compare a group that learns from a game in which words are spoken versus

Table 5.2
The modality principle for games

Source	Game	Feature	Content	Age group	Effect size
Moreno & Mayer (2002a, expt. 1a)	*Design-a-Plant* (desktop)	Spoken words	Botany	College	0.93
Moreno & Mayer (2002a, expt. 1b)	*Design-a-Plant* (sitting head-mounted display [HMD])	Spoken words	Botany	College	0.62
Moreno & Mayer (2002a, expt. 1c)	*Design-a-Plant* (walking HMD)	Spoken words	Botany	College	2.79
Moreno & Mayer (2002a, expt. 2a)	*Design-a-Plant* (desktop)	Spoken words	Botany	College	0.74
Moreno & Mayer (2002a, expt. 1a)	*Design-a-Plant* (walking HMD)	Spoken words	Botany	College	2.24
Moreno et al. (2001, expt. 4a)	*Design-a-Plant* (on-screen agent)	Spoken words	Botany	College	0.60
Moreno et al. (2001, expt. 4b)	*Design-a-Plant* (no on-screen agent)	Spoken words	Botany	College	1.58
Moreno et al. (2001, expt. 5a)	*Design-a-Plant* (on-screen agent)	Spoken words	Botany	College	1.41
Moreno et al. (2001, expt. 5b)	*Design-a-Plant* (no on-screen agent)	Spoken words	Botany	College	1.71
Median					1.41

a group that learns from the same game but with words printed on the screen. Each row provides the source, game description, feature, content, age group, and effect size for a posttest measure of transfer. All the studies involve the *Design-a-Plant* game (Moreno, Mayer, Spires, & Lester, 2001; Moreno & Mayer, 2002a) intended to teach about the environmental science of botany in which first-year college students imagined they traveled to a new planet with certain environmental conditions, and were asked by an on-screen character named Herman-the-Bug to choose the roots, stem, and leaves of a plant that can survive there. Along the way, Herman-the-Bug showed what happens to the plant and explained how plants grow. As a transfer test, students were asked to design plants for a new environment and describe the environment that would best suit a new plant. In each of the nine experimental comparisons, students performed better on the transfer test when the agent's words were spoken rather than printed on the screen, including when the lesson was presented on a desktop computer (Moreno & Mayer, 2002a, experiments 1a and 2a) or immersive virtual reality (Moreno & Mayer, 2002a, experiments 1b, 1c, and 2b), and when the Herman-the-Bug's image was visible on the screen (Moreno et al., 2001, experiments 4a and 5a) or not (Moreno et al., 2001, experiments 4b and 5b).

A limitation of the modality principle for games is that it is based on research with a single game (i.e., *Design-a-Plant*) carried out in a single lab. Although the current evidence is promising, more work is needed to determine whether the modality principle applies to other game environments.

Personalization Principle for Games

The second row in table 5.1 lists the personalization principle for games: people learn better from a game when the words are in conversational style rather than formal style.

The theoretical rationale for the personalization principle is that learners are motivated to exert more effort to make sense of the situation when they feel that the computer is a social partner rather than an uncaring machine. When words are presented in a conversational style, learners are more likely to take a social stance in which they view the computer as a partner. In this way, a conversational style encourages the learner to use any available processing capacity to think more deeply about the underlying nature of the game. Therefore, the personalization principle represents an example of fostering generative processing in the cognitive theory of multimedia learning, as discussed in chapter 3.

The personalization principle was upheld in eight out of eight experimental comparisons, yielding a median effect size of $d = 1.54$, which is a large effect. Overall, this review shows that the personalization principle, which has been identified as a principle for the design of multimedia lessons (Mayer, 2009), also applies to the design of educational games.

Table 5.3 summarizes each of eight experimental comparisons that compare a group that learns from a game in which words are in a conversation style (personalization group) versus a group that learns from the same game with words in a formal style (control group). Each row provides the source, game description, feature, content, age group, and effect size for a posttest measure of transfer. The first two rows summarize experiments by Cordova and Lepper (1996) in which elementary school children played a fantasy math game (e.g., *Space Quest*) aimed at teaching the order of operations for

Table 5.3
The personalization principle for games

Source	Game	Feature	Content	Age group	Effect size
Cordova & Lepper (1996)	Math game (without choice)	Learner's name	Arithmetic	Elementary	1.16
Cordova & Lepper (1996)	Math game (with choice)	Learner's name	Arithmetic	Elementary	1.54
Wang et al. (2008)	*Virtual Factory*	Polite style	Engineering	College	0.93
Moreno & Mayer (2000, expt. 3)	*Design-a-Plant* (on-screen text)	Conversational style	Botany	College	1.55
Moreno & Mayer (2000, expt. 4)	*Design-a-Plant* (narration)	Conversational style	Botany	College	1.58
Moreno & Mayer (2000, expt. 5)	*Design-a-Plant* (narration)	Conversational style	Botany	College	0.89
Moreno & Mayer (2004, expt. 1a)	*Design-a-Plant* (desktop)	Conversational style	Botany	College	1.58
Moreno & Mayer (2004, expt. 1b)	*Design-a-Plant* (virtual reality)	Conversational style	Botany	College	1.93
Median					1.54

arithmetic. Students who played a game that included comments using their name (personalized group) performed better on a subsequent twenty-item posttest involving applying the principle of order operations than did students who received the same comments but without their name or personal information (control group), both when students had some choice about nonessential game elements such as icons and names ($d = 1.54$) and when then did not have choice ($d = 1.16$).

The third row summarizes a study by Wang et al. (2008) in which college students played an industrial engineering simulation game called *Virtual Factory* aimed at teaching how to design assembly lines. For some students, the on-screen agent provided hints and feedback using polite wording (polite group), whereas for other students the on-screen agent used direct wording throughout the game (control group). On a subsequent posttest involving comprehension of the key principles of assembly lines, nonengineering students performed better when given polite wording (d = 0.93). Nevertheless, there was no personalization effect for engineering students who had strong computer skills.

The final five rows in table 5.3 involve *Design-a-Plant* (Moreno & Mayer, 2000, 2004), as described in the previous section on the modality principle. In each of the five experimental comparisons, college students performed better on a transfer test when Herman-the-Bug's words were in a conversational style (e.g., using first- and second-person as well as self-referencing comments) rather than a formal style, including when the lesson was presented on a desktop computer (Moreno & Mayer, 2004, experiment 1a) or immersive virtual reality (Moreno & Mayer, 2004, experiments 1b), and when Herman-the-Bug's words were spoken as narration (Moreno & Mayer, 2000, experiments 4 and 5) or printed on the screen (Moreno & Mayer, 2000, experiment 3).

A possible boundary condition identified in the Wang et al. (2008) study is that the personalization principle may apply most strongly to novices rather than experts in computer technology. Further research is warranted to determine boundary conditions for the personalization principle for games.

Pretraining Principle for Games

The third row in table 5.1 lists the pretraining principle for games: people learn better from a game when they receive pregame experiences for foundational knowledge.

The theoretical rationale for the pretraining principle is that the learner's cognitive system can become overloaded by the need to become acquainted with game elements, thereby limiting the learner's ability to comprehend the essential instructional content. To help reduce this load, pretraining gives learners experience with the key elements and their characteristics, thereby freeing up limited cognitive capacity during subsequent game playing to focus on the relations among them. Thus, the pretraining principle represents an example of managing essential processing in the cognitive theory of multimedia learning, as explained in chapter 3.

The pretraining principle was upheld in seven out of seven experimental comparisons, yielding a median effect size of $d = 0.77$, which is a medium-to-large effect. Overall, this review shows that the pretraining principle,

Table 5.4
The pretraining principle for games

Source	Game	Feature	Content	Age group	Effect size
Leutner (1993, expt. 1)	Farming simulation game	Pregame tutorial	Business	Secondary	0.55
Leutner (1993, expt. 3)	Farming simulation game	Tutorial sheet	Business	Secondary	0.64
Swaak et al. (1998)	*Oscillation*	Begin with lower levels	Physics	College	0.78
de Jong et al. (1999)	*Collisions*	Begin with lower levels	Physics	College	0.91
Mayer et al. (2002, expt. 2)	Geology simulation game	Pretraining with drawings	Geology	College	0.57
Mayer et al. (2002, expt. 3)	Geology simulation game	Pretraining with drawings	Geology	College	0.85
Fiorella & Mayer (2012, expt. 1)	*Circuit Game*	Pretraining in principles and circuit symbols	Electricity	College	0.77
Median					0.77

which has been identified as a principle for the design of multimedia lessons (Mayer, 2009), also applies to the design of educational games.

Table 5.4 summarizes each of seven experimental comparisons between a group that learns from a game that includes pregame experiences aimed at building foundational knowledge (pretraining group) versus a group that learns from the same game without pregame experiences (control group). Each row provides the source, game description, feature, content, age group, and effect size for a posttest measure of transfer. The first two rows summarize experiments by Leutner (1993), in which high school or college students played a farming simulation game by taking the role of a family farmer, and making decisions about how to farm a ten-acre parcel under varying climate and ecological conditions. Some students (pretaining group) were given a brief tutorial on farming before the game (experiment 1) or as a sheet placed beside the computer (experiment 3), whereas other students (control group) received no background information. Across two separate experiments, students in the pretraining group performed much better than students in the base group on a subsequent twenty-seven-item knowledge test that involved applying what was learned.

As summarized in the third row, Swaak, van Joolingen, and de Jong (1998) asked college students to interact with a simulation called *Oscillation* in which they learned about the physics of motion. Some students (control group) were given the highest level, in which many variables could be manipulated at once (level three), whereas other students (pretraining group) first worked on a progression of lower levels (in which fewer variables could be manipulated) before moving on to level three. On a posttest, the pretraining group performed better than the control group on making predictions for what-if problems.

The fourth row summarizes a study by de Jong et al. (1999), in which college students interacted with a simulation called *Collisions* that taught about the physics of motion. Some students (control group) were given the highest two levels (levels four and five) to work on for ninety minutes, whereas other students (pretraining group) first worked on a progression of three lower levels before moving on to levels four and five. On a posttest, the pretraining group performed better than the base group on making predictions for what-if problems.

As summarized in the next two rows, Mayer, Mautone, and Prothero (2002) asked college students to play the *Profile Game*, in which they took the role of geologists trying to discover the geologic formations on a section of a new planet. Some students were given a brief pretraining in which they were shown a sheet with illustrations of the types of formations they were looking for, such as a ridge or trench (pretraining group), whereas other students received no pretraining (control group). The pretraining group performed better than the control group on solving the problems ($d = 0.57$ in experiment 2) as well as on solving problems on a transfer posttest ($d = 0.85$ in experiment 3).

Finally, the last row summarizes a study by Fiorella and Mayer (2012) in which college students played the *Circuit Game*, aimed at helping them learn about electric circuits. Students who received a sheet listing eight principles of electricity before the game and were asked to relate it to their game actions (pretraining group) scored better on an embedded problem-solving transfer test than students who were not given the sheet (control group).

The pretraining principle was supported across a variety of situations. The way that pretraining was implemented, however, differed somewhat across studies, so further work may be needed to more clearly delineate the pretraining feature.

Coaching Principle for Games

The fourth row in table 5.1 lists the coaching principle for games: people learn better from a game when they receive advice or explanations throughout the game.

The theoretical rationale for the coaching principle is that learners may be so overwhelmed by the perceptual features of the game and their own motor activity throughout the game that they fail to discover the underlying instructional content. Yet they are more likely to concentrate on the key instructional message of the game when they receive appropriate advice and explanation in the context of their game activity. By receiving just-in-time advice and explanation, learners are encouraged to focus on the essential instructional information and use it to make sense of their game activity. Hence, the coaching principle represents an example of reducing extraneous processing in the cognitive theory of multimedia learning, as described in chapter 3.

The coaching principle was upheld in six out of seven experimental comparisons, yielding a median effect size of $d = 0.68$, which is a medium-to-large effect. The coaching principle goes beyond basic instructional design principles identified for multimedia learning environments (Mayer, 2009), although various forms of feedback have been shown to be powerful instructional features (Hattie, 2009).

Table 5.5 summarizes each of seven experimental comparisons between a group that learns from a game that includes online advice or explanations during game play (coaching group) versus a group that learns from the same game without online advice or explanations (control group). Each row provides the source, game description, feature, content, age group, and effect size for a posttest measure of transfer.

The first three rows in table 5.5 summarize research by Leutner (1993, experiments 1, 2, and 3) in which high school and college students played a farming simulation game, taking on the role of a family farmer and making decisions about how to farm a ten-acre parcel under varying climate and ecological conditions. For some students, online advice was added to the game in which the students received warnings, corrections, and

Table 5.5
The coaching principle for games

Source	Description	Feature	Content	Age group	Effect size
Leutner (1993, expt. 1)	Farming simulation game	Online advice	Business	Secondary	0.85
Leutner (1993, expt. 2)	Farming simulation game	Online advice	Business	College	0.97
Leutner (1993, expt. 3)	Farming simulation game	Online advice	Business	Secondary	0.63
Van Eck & Dempsey (2002)	Math game (with no competition)	Online advice	Math	Secondary	0.86
Van Eck & Dempsey (2002)	Math game (with competition)	Online advice	Math	Secondary	–0.37
Cameron & Dwyer (2005)	*Heart Attack* quiz game after each item	Receive explanations	Physiology	College	0.57
Mayer & Johnson (2010)	Electric *Circuit Game* after each item	Receive explanations	Electricity	College	0.68
Median					0.68

comments in response to their choices as well as the current state of the farm (coaching group), whereas other students played the simulation game without advice (control group). Across three separate experiments, students in the coaching group performed much better than those in the base group on a subsequent twenty-seven-item knowledge test that involved applying what was learned.

As summarized in the next two rows, Van Eck and Dempsey (2002) asked middle school students to play a math game based on an aunt and uncle's remodeling business that either included or did not include online advice from the aunt and uncle in response to requests for help. On a subsequent math posttest involving a different game context, the advice group performed better than the no-advice group when there was no competition (d = 0.86), but not when there was competition (d = –0.37).

The sixth line in table 5.5 summarizes a study by Cameron and Dwyer (2005) in which college students received an online lesson about the human heart and then played an online quiz game called *Heart Attack*.

In the game, they moved around a board and answered multiple-choice questions about how the heart works aimed at building a score to save a patient from a heart attack. On a two-week-delayed, sixty-item test on the parts and relations in the heart, students who had received explanative feedback throughout the game performed better than students given no feedback ($d = 0.57$).

Finally, the final row of table 5.5 summarizes a study by Mayer and Johnson (2010) in which college students played the *Circuit Game*—a puzzle game intended to teach how electric circuits work. Some students received an on-screen explanation after each major decision (coaching group), whereas other students did not (control group). On a subsequent embedded transfer test involving twenty-five different circuit problems, the coaching group outperformed the control group ($d = 0.68$).

Although the coaching principle was supported across Leutner's farming simulation game and Mayer and Johnson's electric circuit game in which coaching was provided to students, it was not supported in the competition version of Van Eck and Dempsey's math game in which the learner had to ask for help. More research is needed to determine whether online advice and explanations should be optional or automatic. In addition, further work is needed to more clearly delineate the coaching feature and specify which forms of feedback should be included.

Self-Explanation Principle for Games

The fifth row in table 5.1 lists the self-explanation principle for games: people learn better from games when they are asked to give explanations during game play.

The theoretical rationale for the self-explanation principle is that learners may have processing capacity available, but choose not to use it to make sense of the game's instructional content. The deeper cognitive processing needed for meaningful learning, though, is primed when the game includes prompts to explain the game's content. In this way, the learner is encouraged to engage in generative processing that otherwise would not have taken place. The self-explanation principle therefore represents an example of fostering generative processing in the cognitive theory of multimedia learning, as described in chapter 3.

The self-explanation principle was supported in five out of six experimental comparisons, yielding a median effect size of $d = 0.81$, which is

Table 5.6
The self-explanation principle for games

Source	Description	Feature	Content	Age group	Effect size
Lee & Chen (2009)	Frog-leaping game	Self-explanation prompts	Patterns	Secondary	3.26
Mayer & Johnson (2010)	*Circuit Game*	Select explanation from list	Electricity	College	0.91
Johnson & Mayer (2010, expt. 1)	*Circuit Game*	Select explanation from list	Electricity	College	1.20
Johnson & Mayer (2010, expt. 2a)	*Circuit Game*	Select explanation from list	Electricity	College	0.71
Johnson & Mayer (2010, expt. 2b)	*Circuit Game*	Type explanation in box	Electricity	College	–0.06
Fiorella & Mayer (2012, expt. 2)	*Circuit Game*	Write in principle	Electricity	College	0.53
Median					0.81

a large effect. This review suggests that a principle that initially was discovered in the context of students reading text (Fonseca & Chi, 2011) may also apply to the design of educational computer games.

Table 5.6 summarizes each of six experimental comparisons between a group that learns from a game that includes prompts for self-explanation (self-explanation group) versus a group that learns from the same game without self-explanation prompts (control group). Each row provides the source, game description, feature, content, age group, and effect size for a posttest measure of transfer.

In a study by Lee and Chen (2009), reported in the first row of table 5.6, ninth-grade students in Taiwan played a frog-leaping game in which the goal was to move frogs to opposite sides of a line based on puzzle game rules. Some students were given additional specific prompts to explain their moves and monitor their progress (self-explanation group), whereas others received only general prompts to complete the task (control group). The self-explanation group performed better than the general prompt group on reasoning on both a simple and complex problem.

As summarized in the next three rows of table 5.6, Mayer and Johnson (2010) and Johnson and Mayer (2010) asked college students to play a puzzle game intended to teach about how electric circuits work, called the *Circuit Game*. Some students were asked to select an explanation for their solution from an on-screen list (self-explanation group), whereas other students were not (control group). On a subsequent embedded transfer test involving twenty-five different circuit problems, the self-explanation group outperformed the control group across three different experiments (Mayer & Johnson, 2010, experiment 1; Johnson & Mayer, 2010, experiments 1 and 2a). However, as shown in the fifth row of table 5.5, asking students to type in an explanation for their solution into an on-screen text box (self-explanation group) resulted in slightly poorer transfer test performance than the control group (Johnson & Mayer, 2010, experiment 2b). Apparently, a possible boundary condition is that self-explanation prompts work best when they are least intrusive.

Finally, the last row in table 5.6 summarizes a study by Fiorella and Mayer (2012) in which some students were asked to complete a worksheet concerning the principles of electric circuits as they played the *Circuit Game* (self-explanation group), whereas other students did not get a worksheet (control group). Students in the self-explanation group who filled out the sheet accurately scored higher on a transfer test than students in the control group ($d = 0.53$), but students in the self-explanation group who did not fill out the sheet accurately did not show a strong improvement over the control group ($d = 0.08$).

Most of the evidence comes from a single gamelike environment called the *Circuit Game*, so it would be useful to have evidence from other games. Concerning boundary conditions, within the *Circuit Game* studies it appears that self-explanation works best when students select explanations from a list rather than type them in freehand. When asked to write short explanations, self-explanation works best when students generate accurate explanations. Overall, students may need some guidance or scaffolding in the self-explanation process.

What Does Not Work: Two Unpromising Features

A secondary goal of this review is to identify game features that have been shown to not improve learning. The bottom section of table 5.1 lists two unpromising features: immersion and redundancy.

Immersion Principle for Games

The first unpromising feature in table 5.1 is the immersion principle: people do not learn better when a game is rendered in realistic 3-D virtual reality rather than in 2-D.

On the surface, it might be common sense to make games as realistic as possible, but that can be a counterproductive move according to the cognitive theory of multimedia learning. Increased realism, such as in the form of 3-D virtual reality, can add extraneous details to the learner's perceptual field, which compete for the learner's limited processing capacity. To the extent that the learner's cognitive system focuses on processing these extraneous details, capacity is not available to engage in essential processing (i.e., representing the main instructional content) and generative processing (i.e., making sense of it). Thus, the cognitive theory of multimedia learning predicts that changing from desktop 2-D game renderings to 3-D renderings in virtual reality can decrease learning by distracting the learner from engaging in appropriate cognitive processing during learning, and certainly is not expected to help learning. The immersion principle represents an example of the need to reduce extraneous processing in the cognitive theory of multimedia learning, as discussed in chapter 3.

In two out of six experimental comparisons, the immersion group outscored the control group, and in four out of six comparisons, the control group outscored the immersion group, yielding a median effect size of $d = -0.14$, which is a negligible negative effect. Overall, making a game more realistic does not appear to contribute to academic learning for beginners.

The top portion of table 5.7 summarizes each of six experimental comparisons between a group that learned from a game rendered in 3-D immersive virtual reality (immersion group) versus a group that learned from the same game rendered on a desktop computer screen in 2-D (control group). In each study, college students played *Design-a-Plant*, as described in the section on the personalization principle. In one study, the immersion group outperformed the control group on a transfer test when the words were spoken (Moreno & Mayer, 2002a, experiments 1a and 2a), but the immersion group performed worse than the control group when the words were presented as printed text (Moreno & Mayer, 2002a, experiments 1b and 2b). In another study, the control group outperformed the immersion group when spoken words were in both personalized and

Table 5.7
Two unpromising principles for games

Source	Description	Feature	Content	Age group	Effect size
Immersion					
Moreno & Mayer (2002a, expt. 1a)	*Design-a-Plant* (narration)	Virtual reality	Botany	College	0.75
Moreno & Mayer (2002a, expt. 1b)	*Design-a-Plant* (printed text)	Virtual reality	Botany	College	–0.61
Moreno & Mayer (2002a, expt. 2a)	*Design-a-Plant* (narration)	Virtual reality	Botany	College	0.73
Moreno & Mayer (2002a, expt. 2b)	*Design-a-Plant* (printed text)	Virtual reality	Botany	College	–0.19
Moreno & Mayer (2004, expt. 1a)	*Design-a-Plant* (personalized style)	Virtual reality	Botany	College	–0.10
Moreno & Mayer (2004, expt. 1b)	*Design-a-Plant*	Virtual reality (formal style)	Botany	College	–0.63
Median					–0.14
Redundancy					
Moreno & Mayer (2002b, expt. 2a)	*Design-a-Plant*	Spoken + printed words	Botany	College	–0.19
Moreno & Mayer (2002b, expt. 2b)	*Design-a-Plant*	Spoken + printed words	Botany	College	–0.25
Median					–0.22

formal styles (Moreno & Mayer, 2004, experiments 1a and 1b). Overall, there is no strong evidence to support the use of highly realistic renderings when high realism is not needed to help students learn the academic content.

Redundancy Principle for Games

The second unpromising principle shown in table 5.1 is the redundancy principle: people do not learn better when a game includes printed and spoken text together rather than spoken text alone.

On the surface, it might appear that providing both printed and spoken text would be a good way to accommodate a variety of learning strategies, but according to the cognitive theory of multimedia learning, this is not a productive instructional strategy. Offering corresponding narration and printed text can create extraneous cognitive load because learners may waste precious capacity by trying to reconcile the two incoming verbal streams; furthermore, the added words on the screen compete for limited processing capacity in the visual channel, thereby limiting what is learned, as in the modality principle. The solution to these problems is to present words only in spoken form, which is an instance of reducing extraneous processing in the cognitive theory of multimedia learning, as explored in chapter 3.

The bottom portion of table 5.7 summarizes two experimental comparisons involving *Design-a-Plant*, both of which favor presenting narration alone versus narration and corresponding on-screen text (Moreno & Mayer, 2002b). In two out of two experimental comparisons, students performed worse on a transfer test when they played a game with complementary printed and spoken words (redundant group) as opposed to the same game with spoken words alone (control group), yielding a median effect size of $d = -0.22$, which is a negligible negative effect. The redundancy principle for games is consistent with similar findings involving the design of online multimedia lessons (Mayer, 2009). Overall, there is no evidence-based rationale for adding on-screen text to narration for students who are capable of understanding spoken text.

What Has Not Yet Been Shown to Work: Six Not-Yet-Promising Features

Another secondary goal of this review is to identify features that have insufficient evidence to determine whether they are effective in improving learning from games. The middle rows of table 5.1 list six such features: competition, segmenting, image, narrative theme, choice, and learner control.

Competition Principle

The first not-yet-promising principle shown in table 5.1 is the competition principle: students do not necessarily learn better when they compete for a prize.

On the positive side, competition may be seen as a motivating feature that encourages learners to engage in generative processing during learning. When challenged, people tend to work harder to rise to the occasion. On the negative side, competition may create extraneous cognitive processing, when learners focus on frenetic activity rather than reflecting on the underlying point of the game. When stressed, people tend to waste cognitive energy coping with their stress. Therefore, the impact of competition features in a game depends on the degree to which people interpret them as a challenge—which motivates generative processing—or a threat—which creates extraneous processing (DeLeeuw & Mayer, 2011).

In two out of three experimental comparisons, students learned more from a game when they did not compete for a prize (control group) than when they did (competition group), with a median effect size of $d = -0.06$, which is a negligible negative effect.

Table 5.8 lists three experimental comparisons involving competition. In Van Eck and Dempsey (2002), middle school students played a math game based on an aunt and uncle's remodeling business (control group), or played the game against an on-screen computer competitor (competition group). On a subsequent math posttest involving a different game context, the competition group performed better than the no-competition group when no online advice was provided during the game ($d = 0.76$), but not when online advice was supplied ($d = -0.52$). More recently, DeLeeuw and Mayer (2011) asked college students to play the *Circuit Game* (as described in a previous section) either with prizes awarded based on an on-screen scoreboard (competition group), or without the scoreboard and prizes (control group). Overall, the control group slightly outperformed the competition group ($d = -0.06$); for men, the control group strongly outperformed the competition group ($d = -0.54$), and for women, the competition group slightly outperformed the control group ($d = +0.24$). Overall, there does not appear to be strong and consistent evidence to support incorporating competitive features into an educational game.

Segmenting Principle for Games

The second not-yet-promising principle shown in table 5.1 is the segmenting principle: students do not necessarily learn better when the game screen is broken into parts.

Table 5.8
Six not-yet-promising principles for games

Source	Description	Feature	Content	Age group	Effect size
Competition					
Van Eck & Dempsey (2002)	Math game (with no advice)	Computer competitor	Math	Secondary	0.76
Van Eck & Dempsey (2002)	Math game (with advice)	Computer competitor	Math	Secondary	–0.52
DeLeeuw & Mayer (2011)	*Circuit Game*	Prizes based on score	Electricity	College	–0.06
Median					–0.06
Segmenting					
Lee, Plass, & Homer (2006)	Ideal gas law simulation (with complex interface)	Break into two screens	Chemistry	Secondary	0.73
Lee, Plass, & Homer (2006)	Ideal gas law simulation (with optimized interface)	Break into two screens	Chemistry	Secondary	–0.19
Park, Lee, & Kim (2009)	Laws of motion simulation (low prior knowledge)	Break into three segments	Physics	Elementary	0.65
Park, Lee, & Kim (2009)	Laws of motion simulation (high prior knowledge)	Break into three segments	Physics	Elementary	–0.98
Median					0.23
Image					
Moreno et al. (2001, expt. 4a)	*Design-a-Plant* (cartoon, voice)	Agent's image on-screen	Botany	College	–0.50
Moreno et al. (2001, expt. 4b)	*Design-a-Plant* (cartoon, printed text)	Agent's image on-screen	Botany	College	0.22
Moreno et al. (2001, expt. 5a)	*Design-a-Plant* (video, voice)	Agent's image on-screen	Botany	College	0.22
Moreno et al. (2001, expt. 5b)	*Design-a-Plant* (video, printed text)	Agent's image on-screen	Botany	College	0.35
Median					0.22
Narrative theme					
Cordova & Lepper (1996)	Math game	Narrative theme	Arithmetic	Elementary	0.60
Adams et al. (2012)	*Cache 17*	Narrative theme	Devices	College	0.16
Median					0.38

Table 5.8 (continued)

Source	Description	Feature	Content	Age group	Effect size
Choice					
Cordova & Lepper (1996)	Math game (without personalization)	Allow choice of icons	Math	Elementary	0.12
Cordova & Lepper (1996)	Math game (with personalization)	Allow choice of icons	Math	Elementary	0.49
Median					0.30
Learner control					
Swaak & de Jong (2001)	Electric circuit simulation	Learner control of content	Electricity	Secondary	0.19

On the positive side, breaking the screen into parts allows for a simplified rendering within each window, which requires less extraneous processing. On the negative side, having multiple windows available at the same time requires coordination among the windows, which requires more extraneous processing. Overall, breaking a screen into parts can reduce extraneous load (and thereby improve learning) when the windows are processed sequentially, but can increase extraneous load (and thereby hurt learning) when the windows must be processed simultaneously.

In two out of four experimental comparisons, students learned more from a simulation game when it was broken into multiple windows, and in two out of four experimental comparisons they learned less, yielding a median effect size of $d = 0.23$ favoring segmenting, which is a small effect.

Table 5.8 lists four experimental comparisons that involved segmenting. As summarized in the first two lines, Lee, Plass, and Homer (2006) asked middle school students in Korea to solve a problem using a computer-based ideal gas law simulation. For some students, all the relations among all three variables were shown on a single screen (control group), whereas for other students the relations among pairs of variables were shown on two separate screens (segmented group). On a subsequent transfer posttest, the segmented group outperformed the base group on a standard version of the simulation ($d = 0.73$), but not when the screens were optimized to minimize cognitive load ($d = -0.19$). Segmenting appears to work best when the learning situation is complicated.

As summarized in the last two rows, Park, Lee, and Kim (2009) asked fifth graders in Korea to use a simulation to learn about the physics of motion, in which each problem in the simulation was segmented into three parts or was presented as a continuous simulation with many interacting variables. On a subsequent comprehension test, low-prior-knowledge students performed better if they had learned with the segmented simulation ($d = 0.65$), whereas high-prior-knowledge students performed better if they had learned with the continuous simulation ($d = -0.98$). Segmenting appears to work best when the learners are inexperienced and therefore more subject to cognitive overload. Overall, research suggests that segmenting may be effective when the simulation game has the potential to be overloading (i.e., when the interface is complex or the learners are inexperienced), but replication evidence is needed to confirm this trend.

Image Principle for Games

The next not-yet-promising feature in table 5.1 is the image principle: people do not necessarily learn better from a game if a pedagogical agent's image is on the screen.

According to the cognitive theory of multimedia learning, the theoretical rationale for the image principle is that the agent's image can distract the learner, thereby instigating extraneous processing. When cognitive resources are wasted on attending to the agent's image, there are fewer resources to attend to relevant material in the visual channel. In short, the image principle represents an example of the need to reduce extraneous processing based on the cognitive theory of multimedia learning. In contrast, under some circumstances the presence of an agent who behaves like a human could serve to promote a sense of social partnership, which would motivate the learner to work harder to make sense out of the game situation. In short, under some circumstances the image principle could also represent an example of how to foster generative processing.

The third row under the not-yet-promising heading in table 5.1 shows that in three out of four experimental comparisons, a group that played a simulation game containing an agent's image on the screen (image group) performed better on a transfer posttest than a group that played the same game without an on-screen agent's image (control group), yielding a median effect size of $d = 0.22$, which is small effect.

As shown in table 5.8, all the evidence comes from a series of four experimental comparisons in a study by Moreno et al. (2001), in which students played *Design-a-Plant* with the help of an on-screen agent. When the on-screen agent was shown as a talking head of an attractive young man, students performed slightly better on a transfer posttest than when no video was included ($d = 0.22$ and 0.35). When the on-screen agent was a cartoon character named Herman-the-Bug, students performed better when he communicated in printed text ($d = 0.22$) and worse when he spoke ($d = -0.50$) as compared to having no agent image on the screen. Overall, there is not strong and consistent evidence to support placing the agent on the screen, but it would be useful to have replication evidence from other game venues.

Narrative Theme Principle for Games

The next not-yet-promising feature in table 5.1 is the narrative theme principle: people do not necessarily learn better from a game when a strong narrative theme is added.

Although common sense suggests that people will be more involved in a game that has a narrative theme they care about, the cognitive theory of multimedia learning warns against this instructional design practice. The major objection is that embellishing a game with a complicated story line can distract the learner, thereby instigating extraneous cognitive processing during game playing. In short, the call to minimize the role of narrative theme can be seen as an example of reducing extraneous processing in the cognitive theory of multimedia learning. In contrast, when the narrative theme is directly related to the instructional goal, it could serve to direct the learner's attention toward the essential material and foster the learner's interest in reflecting on it.

Table 5.1 shows that in two out of two experimental comparisons, students who played an adventure game with a strong narrative theme (narrative theme group) scored higher on a transfer posttest than students who played the same game without a strong narrative theme (control group), yielding a median effect size of $d = 0.38$, which is a small-to-medium effect.

Table 5.8 shows that the main supporting evidence comes from a study by Cordova and Lepper (1996), in which elementary school children played a math game aimed at teaching the order of operations for arithmetic. Students who played a game that had a fantasy theme (such as

Space Quest or *Treasure Hunt*) performed better on a subsequent posttest ($d = 0.60$) than did students who received the identical game without a fantasy theme (i.e., *Math Game*). In contrast, Adams, Mayer, MacNamara, Koening, and Wainess (2012) asked college students to play a narrative discovery game called *Cache 17* that helped students learn about how electromechanical devices work. Students who played the game with a narrative theme about detectives searching for lost artwork performed slightly better on a transfer posttest than students who played a version without a narrative theme ($d = 0.16$). Overall, there is insufficient evidence to determine whether and when narrative theme can aid academic learning in games.

Choice Principle for Games

The next not-yet-promising feature in table 5.1 is the choice principle: people do not necessarily learn better from a game if they can choose aspects of the game format.

The theoretical rationale for choice is self-determination theory (Ryan & Deci, 2009), which proposes that students work harder to learn when they feel intrinsically motivated. Allowing some degree of choice in the appearance of the game interface can increase intrinsic motivation. This idea is consistent with the idea of fostering generative processing within the cognitive theory of multimedia learning.

In two out of two experimental comparisons, students performed better on transfer tests after playing a game in which they could choose the appearance of the icons before the game (choice group) than when they could not (control group), yielding a median effect size of $d = 0.30$, which is a small-to-medium effect. As summarized in table 5.8, the evidence comes from a study by Cordova and Lepper (1996) involving a math game aimed at teaching the arithmetic order of operations to elementary school students. Students who were allowed to choose the icons for the game (choice group) outperformed students who were not (control group) on a transfer posttest, yielding an effect size of $d = 0.12$ for a standard version of the game and $d = 0.49$ when the game was personalized by using the player's name. Overall, there is not yet sufficient evidence for the choice principle, but replication studies using other game venues would be useful.

Learner Control Principle for Games

The last not-yet-promising feature in table 5.1 is the learning control principle: students do not necessarily learn better when they control the order of levels in the game.

Although learner control may appear to be consistent with the idea of child-centered education, the cognitive theory of multimedia learning recommends against this instructional design feature on the grounds that it can create extraneous processing. The act of having to make decisions can use up cognitive resources that would better be applied to understanding what is going on in the game. Players are likely to make suboptimal choices if they lack appropriate metacognitive skill, as is generally the case for inexperienced learners. In short, predicting the ineffectiveness of learner control is based on the need to reduce extraneous processing within the cognitive theory of multimedia learning. In contrast, proponents argue that learner control can increase student motivation to learn, thereby contributing to the goal of fostering generative processing.

As summarized in table 5.8, in a single experimental comparison reported by Swaak and de Jong (2001), high school students interacted with *CIRCUIT*, a simulation in which they learned about electric circuits. For some students, the order of the levels was fixed in a progression from simple to complex (control group), whereas for other students the order was under their control (learner control group). On a posttest, the learner control group performed slightly better than the control group on making predictions for what-if problems (d = 0.19). This single study, with a negligible effect, does not constitute strong and consistent support for incorporating learner control into games.

Discussion

Practical Contributions

As summarized in table 5.9, this review yields five evidence-based principles for game design: modality (d = 1.41), personalization (d = 1.54), pretraining (d = 0.75), coaching (d = 0.68), and self-explanation (d = 0.75). This review also yields two principles that have been shown not to be effective in games: immersion (d = –0.14) and redundancy (d = –0.23). Research is too inconsistent or preliminary concerning six emerging principles:

Table 5.9
Seven evidence-based design principles for instructional games

Principle	Full description
What works: Promising features	
Modality	People learn better from a game when the words are spoken rather than printed
Personalization	People learn better from a game when the words are in conversational style rather than formal style
Pretraining	People learn better from a game when they receive pregame experiences for foundational knowledge
Coaching	People learn better from a game when they receive advice or explanations throughout the game
Self-explanation	People learn better from a game when they are asked to give explanations during game play
What does not work: Unpromising features	
Immersion	People do not learn better when a game is rendered in realistic 3-D virtual reality rather than 2-D
Redundancy	People do not learn better when a game includes printed and spoken text together rather than spoken text alone

competition, segmenting, image, choice, narrative theme, and learner control. Important boundary conditions include the finding that several principles are most effective when the learners are inexperienced, the game environment is otherwise complicated, and there are not other helpful instructional features.

Theoretical Contributions

The cognitive theory of multimedia learning suggests three goals of instructional design: reducing extraneous processing, managing essential processing, and fostering generative processing. The instructional goal of reducing extraneous processing may be achieved through the coaching principle. The instructional goal of managing essential processing may be achieved through the modality and pretraining principles. Finally, the goal of fostering generative processing may be achieved through the personalization and self-explanation principles. The ineffectiveness of the immersion and redundancy principles can be explained in terms of the need to reduce extraneous processing.

Methodological Contributions

This review demonstrates the usefulness of the value-added approach to game research, which compares the learning outcome score of a group that plays a base version of a game (control group) versus the learning outcome score of a group that plays the same game with an instructional feature added (experimental group). It also demonstrates the usefulness of conducting a literature review that focuses solely on value-added research rather than on all genres of game research, when the goal is to identify effective instructional design features for games.

Future Contributions

This review shows that modest progress is being made in the scientific study of how to design effective games for learning, including the identification of five promising features that improve learning with games. Clearly, the scientific study of the instructional design of games is in its early phase, so I offer the following suggestions for potentially useful next steps concerning what is needed and what is not needed.

What is needed includes:

- Better acceptance of the role of scientifically rigorous research on game effectiveness. Although game purists may resent the intrusion of science into what has been considered an artistic enterprise aimed at entertainment, this review demonstrates the contributions of an evidence-based approach to game design for improving education.
- Better acceptance of the crucial role of replication in game research. Although some journal editors prefer to reject work that is not "new," replication is an essential requirement of scientific research in education (Shavelson & Towne, 2002). This review shows that the development of evidence-based principles for game design relies on an evidence base of replication studies—that is, a collection of studies all testing the same game feature in different venues.
- Better appreciation of the value of discovering what does not work. In addition to identifying five game features that appear to have promise for improving game design, this review also shows that it is useful to find convincing evidence for what doesn't work.

- Better understanding of the boundary conditions of game design principles. As the database increases, it is important to pinpoint when and for whom each instructional features is most effective.
- Better evidence-based theories of how people learn from games. As the database increases, the underlying cognitive theories of learning can be enhanced to account for the role of motivation and affect in learning.

What is not needed includes:

- More opinion pieces about how games should revolutionize education. In searching the literature, I found an overabundance of books, chapters, and articles that make strong claims based on weak (or no) evidence. A more useful way to approach improving education would be to identify evidence-based principles for how to design effective computer games.
- More methodologically inadequate studies. In searching the literature, I had to discard many empirical studies because they failed to meet the minimal requirements of an experimental comparison. Concerning experimental design, too many studies failed to include a control group and instead merely showed a pretest-to-posttest gain only for the experimental group. In terms of experimental procedure, too many studies had inadequate sample sizes, sometimes with fewer than ten participants per cell, or failed to enforce random assignment, and instead allowed students to self-select their treatment group. Regarding independent variables, too many studies failed to have an appropriate control group that was identical to the experimental group except for one game feature. Concerning the dependent measure, too many studies failed to measure learning outcome and instead measured everything else ranging from self-reported feelings about learning to the number of keystrokes. Although observational studies have a place in our field, experimental comparisons offer a particularly efficient methodology for testing claims about the effectiveness of instructional interventions (Shavelson & Towne, 2002).

Critics of scientific research on improving the educational effectiveness of games may wish to reject this review on the grounds that researchers are not studying real games that people really play. Indeed, most of the reviewed research involves short episodes of game play with games focused on specific academic goals often implemented within a contrived environment. In contrast, in the real world of gaming players choose to spend hours of their time engaged intensely in individual and multiplayer games.

Indeed, there appears to be two worlds of gaming: the play world, in which the goal is to design games for entertainment without the benefit of scientific research, and the educational world, in which the goal is to understand how to design games for learning based on scientific research. Although the play world of gaming has been fabulously successful in terms of garnering the strong participation of players seeking fun, the educational world of gaming is just beginning to develop an evidence-based focus for learners seeking new knowledge. Overall, this review shows that a scientific approach has something useful to contribute when the goal is to design effective games for learning.

References

Note: Asterisk (*) indicates study is in review database.

*Adams, D. M., Mayer, R. E., MacNamara, A., Koening, A., & Wainess, R. (2012). Narrative games for learning: Testing the discovery and narrative hypothesis. *Journal of Educational Psychology, 104*, 235–249.

Anderson, L. W., Krathwohl, D. R., Airasian, P. W., Cruikshank, K. A., Mayer, R. E., Pintrich, P. R., et al. (2001). *A taxonomy for learning, teaching, and assessing*. New York: Longman.

*Cameron, B., & Dwyer, F. (2005). The effect of online gaming, cognition, and feedback type in facilitating delayed achievement of different learning objectives. *Journal of Interactive Learning Research, 16*, 243–258.

Clark, R. E., Yates, K., Early, S., & Moulton, K. (2011). An analysis of the failure of electronic media and discovery-based learning: Evidence for the performance benefits of guided learning methods. In K. H. Silber & W. R. Foshay (Eds.), *Handbook of improving performance in the workplace* (pp. 263–297). San Francisco: Pfeiffer.

Cohen, J. (1988). *Statistical power analysis for the behavioral sciences* (2nd ed.). Hillsdale, NJ: Erlbaum.

Connolly, T. M., Boyle, E. A., MacArthur, E., Hainey, T., & Boyle, J. M. (2012). A systematic review of empirical evidence on computer games and serious games. *Computers & Education, 59*, 661–686.

*Cordova, D. I., & Lepper, M. R. (1996). Intrinsic motivation and the process of learning: Beneficial effects of contextualization, personalization, and choice. *Journal of Educational Psychology, 88*, 715–730.

*de Jong, T., Martin, E., Zamarro, J., Esquembre, F., Swaak, J., & van Joolingen, W. R. (1999). The integration of computer simulation and learning support: An example from the physics domain of collisions. *Journal of Research in Science Teaching, 36,* 597–615.

*DeLeeuw, K. E., & Mayer, R. E. (2011). Cognitive consequences of making computer-based activities more game-like. *Computers in Human Behavior, 27,* 2011–2016.

Ellis, P. D. (2010). *The essential guide to effect sizes.* New York: Cambridge University Press.

*Fiorella, L., & Mayer, R. E. (2012). Paper-based aids for learning with a computer-based game. *Journal of Educational Psychology, 104,* 1074–1082.

Fonseca, B. A., & Chi, M.T.H. (2011). Instruction based on self-explanation. In R. E. Mayer & P. A. Alexander (Eds.), *Handbook of research on learning and instruction* (pp. 296–321). New York: Routledge.

Gee, J. P. (2007). *Good video games and good learning.* New York: Peter Lang.

Ginns, P. (2005). Meta-analysis of the modality effect. *Learning and Instruction, 15,* 313–332.

Hannafin, R. D., & Vermillion, J. R. (2008). Technology in the classroom. In T. L. Good (Ed.), *Twenty-first-century education: A reference handbook* (Vol. 2; pp. 209–218). Thousand Oaks, CA: SAGE.

Hattie, J. (2009). *Visible learning.* New York: Routledge.

Hayes, R. T. (2005). *The effectiveness of instructional games: A literature review and discussion.* (Technical Report 2005–004). Orlando, FL: Naval Air Warfare Center Training Systems Division.

Honey, M., & Hilton, M. (Eds.). (2011). *Learning science through computer games and simulations.* Washington, DC: National Academy Press.

*Johnson, C. I., & Mayer, R. E. (2010). Adding the self-explanation principle to multimedia learning in a computer-based game-like environment. *Computers in Human Behavior, 26,* 1246–1252.

*Lee, C., & Chen, M. (2009). A computer game as a context for non-routine mathematical problem solving: The effects of type of question prompt and level of prior knowledge. *Computers & Education, 52,* 530–542.

*Lee, H., Plass, J. L., & Homer, B. D. (2006). Optimizing cognitive load for learning from computer-based science simulations. *Journal of Educational Psychology, 98,* 902–913.

*Leutner, D. (1993). Guided discovery learning with computer-based simulation games: Effects of adaptive and non-adaptive instructional support. *Learning and Instruction, 3,* 113–132.

Lipsey, M. W., & Wilson, D. B. (2001). *Practical meta-analysis*. Thousand Oaks, CA: Sage.

Low, R., & Sweller, J. (2005). The modality principle in multimedia learning. In R. E. Mayer (Ed.), *The Cambridge handbook of multimedia learning* (pp. 147–158). New York: Cambridge University Press.

Mayer, R. E. (2005). Cognitive theory of multimedia learning. In R. E. Mayer (Ed.), *The Cambridge handbook of multimedia learning* (pp. 31–48). New York: Cambridge University Press.

Mayer, R. E. (2009). *Multimedia learning* (2nd ed.). New York: Cambridge University Press.

Mayer, R. E. (2011). Multimedia learning and games. In S. Tobias & J. D. Fletcher (Eds.), *Computer games and instruction* (pp. 281–306). Amsterdam: Elsevier.

*Mayer, R. E., & Johnson, C. I. (2010). Adding instructional features that promote learning in a game-like environment. *Journal of Educational Computing Research, 42*, 241–265.

*Mayer, R. E., Mautone, P. D., & Prothero, W. (2002). Pictorial aids for learning by doing in a multimedia geology simulation game. *Journal of Educational Psychology, 94*, 171–185.

McGonigal, J. (2011). *Reality is broken: How games make us better and they can change the world*. New York: Penguin Press.

*Moreno, R., & Mayer, R. E. (2000). Engaging students in active learning: The case for personalized multimedia messages. *Journal of Educational Psychology, 92*, 724–733.

*Moreno, R., & Mayer, R. E. (2002a). Learning science in virtual reality environments: Role of methods and media. *Journal of Educational Psychology, 94*, 598–610.

*Moreno, R., & Mayer, R. E. (2002b). Verbal redundancy in multimedia learning: When reading helps listening. *Journal of Educational Psychology, 94*, 156–163.

*Moreno, R., & Mayer, R. E. (2004). Personalized messages that promote science learning in virtual environments. *Journal of Educational Psychology, 96*, 165–173.

*Moreno, R., Mayer, R. E., Spires, H. A., & Lester, J. (2001). The case for social agency in computer-based teaching: Do students learn more deeply when they interact with animated pedagogical agents? *Cognition and Instruction, 19*, 177–213.

Mosteller, F., & Boruch, R. (Eds.). (2002). *Evidence matters: Randomized trails in education research*. Washington, DC: Brookings Institution Press.

O'Neil, H. F., & Perez, R. S. (Eds.). (2008). *Computer games and team and individual learning*. Amsterdam: Elsevier.

O'Neil, H. F., Wainess, R., & Baker, E. L. (2005). Classification of learning outcomes: Evidence from the computer games literature. *Curriculum Journal, 16,* 455–474.

*Park, S. I., Lee, G., & Kim, M. (2009). Do students benefit from interactive computer simulations regardless of prior knowledge levels? *Computers & Education, 52,* 649–655.

Phye, G. D., Robinson, D. H., & Levin, J. (Eds.). (2005). *Empirical methods for evaluating educational interventions.* San Diego: Academic Press.

Prensky, M. (2006). *Don't bother me mom—I'm learning.* Saint Paul, MN: Paragon House.

Randel, J. M., Morris, B. A., Wetzel, C. D., & Whitehill, B. V. (1992). The effectiveness of games for educational purposes: A review of recent research. *Simulation and Games, 23,* 261–276.

Rittenfeld, U., & Weber, R. (2006). Video games for entertainment and education. In P. Vorderer & J. Bryant (Eds.), *Playing video games* (pp. 399–414). Mahwah, NJ: Erlbaum.

Rosenthal, R., Rosnow, R. L., & Rubin, D. B. (2000). *Contrasts and effect sizes in behavioral research.* New York: Cambridge University Press.

Ryan, R. M., & Deci, E. (2009). Self-worth theory: Retrospective and prospects. In K. R. Wentzel & A. Wigfield (Eds.), *Handbook of motivation at school* (pp. 141–170). New York: Routledge.

Shaffer, D. W. (2006). *How computer games help children learn.* New York: Palgrave Macmillan.

Shavelson, R. J., & Towne, L. (Eds.). (2002). *Scientific research in education.* Washington, DC: National Academy Press.

*Swaak, J., & de Jong, T. (2001). Learner versus system control in using online support for simulation-based discovery learning. *Learning Environments Research, 4,* 217–241.

*Swaak, J., van Joolingen, W. R., & de Jong, T. (1998). Supporting simulation-based learning: The effects of model progression and assignments on definitional and intuitive knowledge. *Learning and Instruction, 8,* 235–252.

Sweller, J. (2005). Implications of cognitive load theory for multimedia learning. In R. E. Mayer (Ed.), *The Cambridge handbook of multimedia learning* (pp. 19–30). New York: Cambridge University Press.

Sweller, J., Ayres, P., & Kalyuga, S. (2011). *Cognitive load theory.* New York: Springer.

Tobias, S., & Fletcher, J. D. (2012). Reflections on "a review of trends in serious gaming." *Review of Educational Research, 82*, 233–237.

Tobias, S., Fletcher, J. D., Dai, D. Y., & Wind, A. P. (2011). Review of research on computer games. In S. Tobias & J. D. Fletcher (Eds.), *Computer games and instruction* (pp. 525–545). Charlotte, NC: Information Age Publishing.

*Van Eck, R., & Dempsey, J. (2002). The effect of competition and contextualized advisement on the transfer of mathematics skills in a computer-based instructional simulation game. *Educational Technology Research and Development, 50*, 23–41.

Vogel, J. F., Vogel, D. S., Cannon-Bowers, J., Bowers, C. A., Muse, K., & Wright, M. (2006). Computer gaming and interactive simulations for learning: A meta-analysis. *Journal of Educational Computing Research, 34*, 229–243.

*Wang, N., Johnson, W. L., Mayer, R. E., Rizzo, P., Shaw, E., & Collins, H. (2008). The politeness effect: Pedagogical agents and learning outcomes. *International Journal of Human-Computer Studies, 66*(2): 98–112.

Young, M. F., Slota, S., Cutter, A. B., Jalette, G., Mullin, G., Lai, B., et al. (2012). Our princess is in another castle: A review of trends in serious gaming for education. *Review of Educational Research, 82*, 61–89.

6 Cognitive Consequences Approach: What Is Learned from Playing a Game?

Deanne M. Adams and Richard E. Mayer

Chapter Outline

Summary

The cognitive consequences approach to game research compares the cognitive skill performance of students who are assigned to play an off-the-shelf computer game for an extended period (game group) to those who engage in an alternative activity (control group). A review of the cognitive consequences game research literature identified one promising type of cognitive skill that may be improved through playing one kind of game: playing first-person shooter games (such as *Unreal Tournament* or *Medal of Honor*) can improve perceptual attention skills (d = 1.18 based on eighteen comparisons). There is also preliminary evidence that playing spatial puzzle games (such as *Tetris*) can improve 2-D mental rotation of *Tetris* and *Tetris*-like shapes (d = 0.82 based on six comparisons). Finally, there is a hint that some spatial games may affect 3-D mental rotation and that brain-training games may affect executive function skills, but a larger and more consistent database is needed. The effects generally appear to be strongest for games that require players to repeatedly engage in a specific targeted skill and receive continuous feedback. The vast majority of studies did not find strong positive effects for game playing. There was no pattern of positive effects of game playing on reasoning, motor, and memory tasks. The results support the specific transfer of general skill hypothesis in which playing a particular game can improve student performance on applying the cognitive skills tapped by the game. Overall, the cognitive consequences approach helps determine the degree to which playing a specific kind of game can affect changes in a specific kind of cognitive skill.

Introduction

Rationale for the Cognitive Consequences Approach to Game Research

Suppose you download a new game called *Blast* onto your smartphone (or tablet or laptop computer). The creators of *Blast* claim that playing the game will improve your memory and reasoning skills. How could you test that claim?

One approach could be to play the game every day for a week and see whether your performance improves. Yet simply getting better at the game does not necessarily mean you are improving on important cognitive skills, so game improvement alone does not allow us to adequately test the claim.

Another approach could be to take a pretest on memory and reasoning skills, and then retake it after playing *Blast* for a week to see if you show an improvement. However, the problem with this approach is that you could have improved on the posttest simply because you already took the test before, so only knowing about the pretest-to-posttest gains for game players does not allow us to adequately test the claim.

In contrast to these scientifically questionable approaches, the approach advocated in this book is to conduct cognitive consequences research. In the cognitive consequences approach, you compare the pretest-to-posttest gains (or posttest score) on memory and reasoning tasks of a group that has been assigned to play *Blast* over an extended period against the gains (or posttest scores) of a group that has been assigned to do something other than play the game. As described in chapter 2, cognitive consequences research is designed to test causal claims such as the claim that playing *Blast* will increase your memory and reasoning skills. If you are interested in taking a scientific approach to testing the claims for *Blast* (or other computer games), the cognitive consequences approach is for you.

From the start of the video game revolution in the 1970s and 1980s, hopes were high that playing video games could have positive effects for game players. For example, in an early book on the psychology of video games titled *Mind at Play*, Loftus and Loftus (1983) wondered out loud whether video games could become learning devices: "It would be comforting to know that the seemingly endless hours young people spend playing Defender and Pac-Man were really teaching them something useful. Is the massive amount of practice in motor coordination going to be useful for anything besides being a better video game player? As relevant research has not yet been performed, we can only speculate about direct educational benefits" (p. 121).

If we fast-forward thirty years, the names of the games have changed, and their sophistication and availability have improved, but the popularity of game playing has persisted for a generation, and the cognitive consequences question raised by Loftus and Loftus remains an important one: What do people learn from playing a video game over an extended period?

Today, however, we are beginning to see a healthy and growing research base on the cognitive consequences of game playing that can help us determine whether people learn "something useful" from playing off-the-shelf video games (even when the shelves are in online app stores).

In particular, cognitive consequences research has examined whether perceptional attention skills, mental rotation skills, other spatial cognition skills, and executive function skills can be improved by playing games that range from puzzle games like *Tetris* to first-person shooter games such as *Unreal Tournament 2004*. Furthermore, some preliminary research investigates whether game playing affects reasoning, motor, and memory skills. Table 6.1 provides a definition and example of each of these kinds of cognitive

Table 6.1
What skills may be learned from playing computer games?

Skill	Description	Example
Perceptual attention	Ability to rapidly register and track visual elements	Tell the direction of an object that is flashed on the screen at various eccentricities, or track multiple moving objects on the screen
2-D mental rotation	Ability to mentally rotate a 2-D image	See two 2-D objects with one rotated and either flipped or not flipped; tell whether the two shapes are the same
3-D mental rotation	Ability to mentally rotate a 3-D image	See two 3-D objects with one rotated and either flipped or not flipped; tell whether the two shapes are the same
Spatial cognition	Ability to manipulate mental images and visualize changes in them	See a set of shapes and determine how they fit together into a rectangle
Executive function	Ability to control cognitive processing	Name the ink color of color words that are printed in various colors while suppressing the automatic reading response
Reasoning	Ability to make inferences	Given a series of shapes that change based on a rule, predict what will come next
Motor	Ability to construct or handle something with one's own hands	Given a collection of metal pieces, use your left hand to fit them into a container
Memory	Ability to recall what was presented	Hear a list of digits and repeat them back

skills. The goal of this review is to examine the evidence that playing a game can cause cognitive changes in the player's cognitive skills, particularly the ones listed in table 6.1.

Cognitive consequences research is motivated by several different practical considerations, including improving education and counteracting cognitive aging. First, some cognitive consequences research is motivated in part by the need to expand student participation in academic fields, including encouraging the participation of underrepresented groups such as women and minorities. A recent report by the National Science Foundation (2010), makes the case that to be a leader in research and development, nations need to attract students into science, technology, engineering, and mathematics (STEM) fields. Thus, an important issue is whether game playing can help students develop the skills they need for academic success.

Uttal and Cohen (2012) propose that spatial ability actually acts as a gateway to getting into STEM fields. Experts develop contextualized spatial abilities and specialized semantic knowledge, but novices must rely on their decontextualized spatial abilities. Otherwise-qualified students may be pushed away from STEM fields because their underdeveloped spatial skills impede initial learning. For example, research by Wai, Lubinski, and Benbow (2009) found connections between spatial ability and later success in STEM areas. A consensus report from the National Research Council (2006) titled *Learning to Think Spatially* contended that "spatial thinking is integral to the everyday work of scientists and engineers," and "spatial thinking is a skill that can—and should—be learned by everyone" (p. 5). One intriguing idea explored in this chapter is that playing computer games may be a way to help students develop the cognitive skills they need for learning in academic disciplines, including spatial skills for STEM fields.

An encouraging study examining the link between game playing and success in a STEM field comes from Sanchez (2012), in which playing a game that placed demand on spatial skills facilitated science learning. Students either played the action first-person shooter game *Halo: Combat Evolved* or a word game called *Word Whomp* for twenty-five minutes before reading a science text on plate tectonics. To assess how much they learned, students were then asked to write an essay on "What caused Mt. St. Helens to erupt?" The results showed that students in the action game condition performed significantly better on the essay task although there were no significant differences in prior knowledge between the two groups. In addition, the participants in the action first-person shooter game group also demonstrated significantly greater improvement on a mental rotation task. These results suggest that playing an action video game can prime spatial skills, which in turn can facilitate learning in academic areas that require those abilities. In short, this preliminary work encourages the hope that research on the cognitive consequences of game playing ultimately can pinpoint ways to help students acquire cognitive skills that will increase their chances of success in various academic fields, including STEM fields. The first step in this process is to assess the degree to which game playing can improve cognitive skills, which is the issue addressed in this chapter.

Another practical goal of some cognitive consequences research is to improve skills that may have decreased as a result of cognitive decline due to aging. Whitlock, McLaughlin, and Allaire (2012) and Basak, Boot, Voss, and Kramer (2008) both examined how interventions using games could improve cognitive functioning in areas such as executive control and visual attention in older populations. It is important to note that while these game interventions may not be effective in certain age groups such as college students, older or younger age groups may benefit either by developing new skills (De Lisi & Wolford, 2002) or helping to train skills that may have declined over time.

Theoretical Framework

Cognitive consequences researchers are examining which cognitive skills may be improved through game playing, such as those listed in table 6.1. Anderson and Bavelier (2011) argue that video games that place high demands on certain processes should facilitate increases in those aspects of cognition; therefore, a match between the skills taxed during the game and those required on a test should result in positive effects of game playing. For example, playing fast paced, perceptually demanding first-person shooter games should increase performance on nongame tasks that require perceptual attention processing (such as distinguishing among target objects in a complex visual array).

If table 6.1 represents a list of to-be-enhanced skills, what are the games that can be used to do the job? As an example, figure 6.1 presents a screenshot from the spatial puzzle game *Tetris*, in which the goal is to press

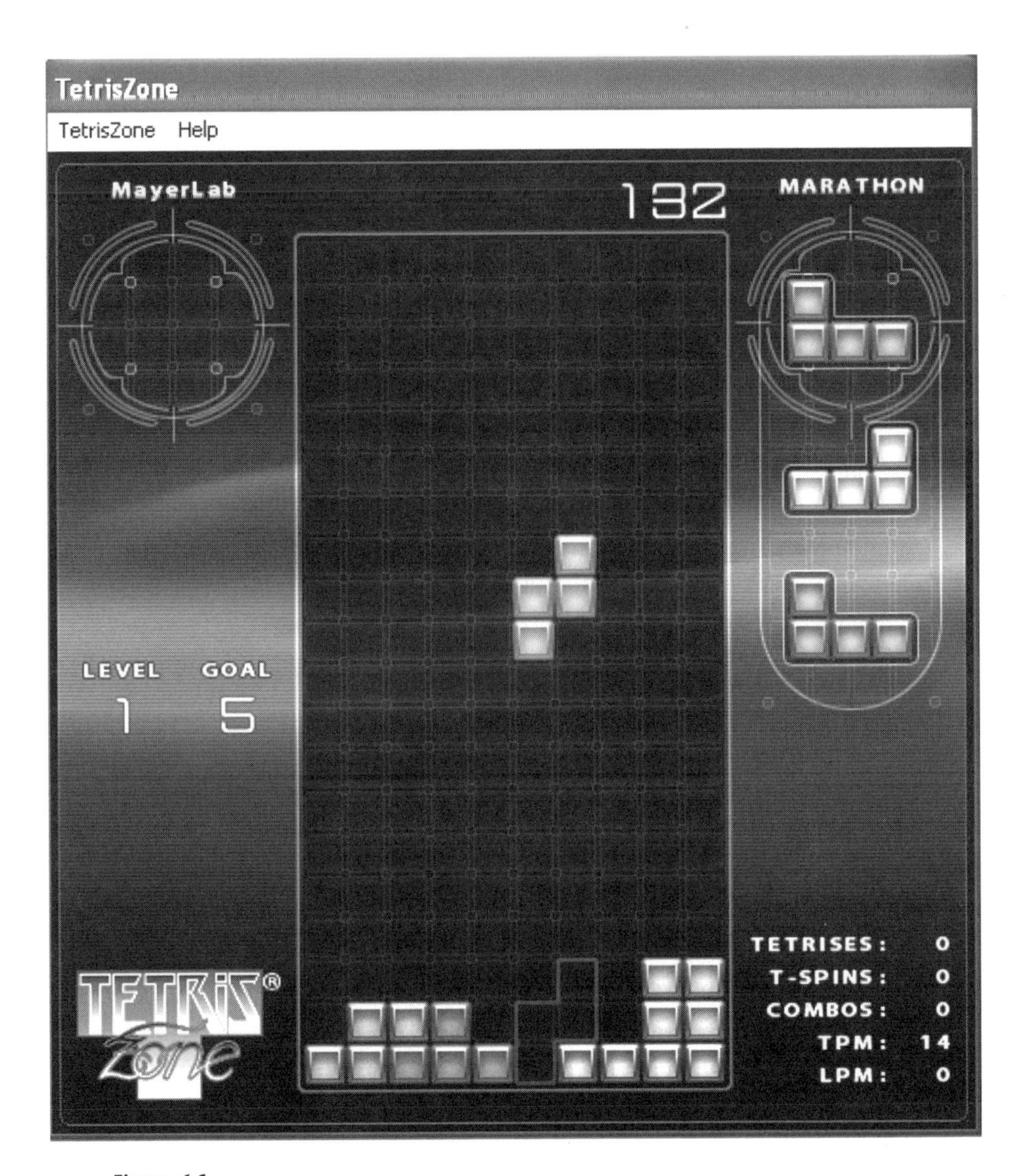

Figure 6.1
Screenshot from *Tetris*

buttons on the controller (or keyboard) to rotate and move shapes that fall from the top of the play area in order to create unbroken lines of blocks by filling in empty spaces. Every time a line is completed, the line disappears and the player is awarded points. As the player's score increases, the falling rate of the shapes increases, making the game harder, and the game ends when incomplete lines pile up to the top of the play area. *Tetris* is the most studied spatial puzzle game (De Lisi &Wolford, 2002; Okagaki & Frensch, 1994; Terlecki, Newcombe, & Little 2008).

What are the cognitive skills required to play *Tetris*? An obvious candidate is the spatial skill of mental rotation. Although the game appears to require mental rotation, Kirsh and Maglio (1994) have shown that players can off-load cognitive processing by using the game mechanics (i.e., by pressing buttons on the keyboard or screen) to physically rotate and move the shapes more efficiently than actually engaging in mental rotation. The game may require other cognitive skills such as visualizing all shape configurations or planning ahead by arranging the falling shapes in anticipation of particular shapes that will create multiple lines, resulting in winning more points.

As another example, figure 6.2 presents a screenshot from the first-person shooter game *Unreal Tournament 2004*, in which players must navigate a two-story, 3-D building environment, while keeping track of the spawning positions of guns and ammunition along with trying to stay alive while fighting against enemy "bots" controlled by the game or other players. In order to be successful, the players must be able to avoid enemy fire while aiming and returning fire, so the game keeps track of how many times the player kills an enemy as well as how many times the player is killed. *Unreal Tournament 2004* (which we refer to simply as *Unreal Tournament*) has been used repeatedly in video game research (Green & Bavelier 2006a, 2006b, 2007; Li, Polat, Makous, & Bavelier, 2009).

What are the cognitive skills required to play *Unreal Tournament*? Players must track moving objects and predict their trajectory at various points in the future, in a rapidly changing environment. Players must also quickly react to the appearance of new visual information, such as the appearance of enemies, so fast but accurate reactions to target objects are important. Thus, it appears that *Unreal Tournament* requires a variety of perceptual attention skills.

Figure 6.2
Screenshot from *Unreal Tournament*

Overall, theoretical progress can be made to the extent that researchers can pinpoint the cognitive skills tapped by a game, and then measure changes in that skill and other cognitive skills due to game playing. In particular, we explore three competing hypotheses concerning the cognitive consequences of playing computer games based on cognitive theories of transfer (Mayer & Wittrock, 2006; Sims & Mayer, 2002; Singley & Anderson, 1989):

Specific transfer Game playing improves skills that are identical to those required to play the game (e.g., playing *Tetris* will only improve the mental rotation of *Tetris* shapes because the game requires repeated mental rotation of *Tetris* shapes). This view of transfer suggests that game playing will not have a major impact on students' cognitive skills unless the games are targeted to specific content. Support for this hypothesis would be reflected in finding that game playing does not have a strong effect on improving cognitive skills required for academic learning.

General transfer Game playing improves the mind in general and therefore improves a wide variety of cognitive skills beyond those targeted in the game (e.g., playing *Tetris* will improve a variety of spatial and perceptual skills even though they are not specifically required to play the game). This view of transfer suggests that game playing will result in general improvements across a variety of cognitive skills. Support for this hypothesis would be reflected in the finding that playing a well-designed computer game has strong effects on improving an array of cognitive skills required for academic learning.

Specific transfer of general skills Game playing allows players to perfect cognitive skills required by game play and apply them to new situations (e.g., playing *Tetris* will improve the mental rotation of various kinds of shapes, but will not affect unrelated cognitive skills). This view of transfer suggests that game playing will result in improvements in the cognitive skills that are required in the game and can be applied to corresponding academic content. Support for this hypothesis would be reflected in the finding that playing a particular game has strong effects on improving cognitive skills tapped by the game and being able to apply them in academic contexts. This is the view proposed by Anderson and Bavelier (2011) in this section, and the one that receives some support in this chapter.

Method

The primary goal of this cognitive consequences review is to summarize the present state of the research on how playing computer games can affect improvements in various cognitive skills. In this review, we focus on research involving novices or nongame players. Although there is also a body of work that examines cognitive differences between regular video game players (VGPs) and non-video-game players (NVGPs), the goal of this review is to examine how playing a game can affect the cognitive skills of individuals who are not regularly exposed to game playing.

Evidence Collection

In searching for papers that fit our cognitive consequences criteria, we examined online databases, including PsychINFO, ERIC, and PubMed, using keywords such as "video games," "games and spatial cognition," "action games," "games and cognition," and "games." We also reviewed the references from previous literature reviews (e.g., Anderson & Bavelier, 2011; Boot, Blakely, & Simons, 2011; Cohen, Green, & Bavelier, 2008; Spence & Feng, 2010). In addition, we conducted a search of all the "cited by" papers for prominent papers in the field, including Green and Bavelier (2003), Okagaki and Frensch (1994), and Sims and Mayer (2002).

Evidence Selection

When determining whether a paper would be included in the review, we used the following criteria:

- The independent variable involves a comparison between a group that has played a game aimed at increasing a particular skill (game group) versus either a group that has not played any game or a group that has played a game that is theorized to not cause changes in the particular cognitive skill (control group). The study must include a control group in order to ensure that the changes in the cognitive skill were due to playing the game and not a simple retest effect. In this way, we meet the criteria of experimental control and random assignment, as discussed in chapter 2.
- The video game used as the independent variable is a commercially available product, and was not developed mainly to teach particular content such as subject material for mathematics, reading, or science.
- At least one of the dependent measures involves a perceptual attention skill, mental rotation skill, spatial cognition skill (other than mental rotation), executive function skill, reasoning skill, motor skill, or memory skill. This requirement addresses the need for appropriate measurement, as described in chapter 2.
- Participants in the study were novice game players for both conditions.
- The paper reports the mean (*M*) and standard deviation (*SD*) for a measure of performance gain (or posttest performance) and sample size (*n*) for each group, or contains other statistical information that allows for computing the value of effect size based on Cohen's *d*.

- The paper is published in a peer-reviewed research journal or book that is accessible.

Evidence Coding

During the evidence-coding phase, we created a database for each experiment that featured a comparison between a game group and a control group. For each comparison, we coded for the citation source, name of game, type of game, comparison group, game-playing time, name of test, type of test, age category, and effect size. In coding for the type of game employed during the manipulation, the game categories included puzzle games (e.g., *Tetris*), first-person shooter games (e.g., *Unreal Tournament* or *Medal of Honor*), real-time strategy games (*Red Alert 2* or *Rise of Nations*), brain-training games (e.g., *Brain Age*), and racing games (e.g., *Antz Extreme Racing*). In coding for the type of test, we sorted data into the cognitive skills categories of perceptual and attention skills, 2-D mental rotation , 3-D mental rotation, spatial cognition skills (other than mental rotation), executive function skills, reasoning skills, motor skills, and memory skills, as summarized in table 6.1. For studies that included multiple dependent measures that tested a variety of skills, multiple entries were created representing each cognitive skill measure whenever a specific effect size could be calculated. The age categories were elementary school, secondary school, college, and older adults.

Evidence Summarization and Interpretation

During the evidence summarization phase we computed effect size for each experimental comparison using Cohen's (1988) *d*, as described in the previous chapter. We partitioned the analysis based on the type of cognitive skill being tested—perceptual attention, 2-D mental rotation, 3-D mental rotation, spatial cognition, executive function, reasoning, motor, and memory skills (represented in tables 6.3 through 6.10). For each target skill, we partitioned the comparisons within each table by the type of game, such as first-person shooter, spatial puzzle, real-time strategy, racing, and brain-training games. We computed the mean effect sizes to determine the effects of playing a particular type of game on a particular type of cognitive skill. For target skills involving an insufficient number of studies for this level of analysis, we combined all games together to compute the median effect size of game playing on the target skill. The results section is broken into subsections that summarize the effects of playing a particular type of game on a particular type of cognitive skill.

Results

Table 6.2 summarizes the effects of game playing on each of the cognitive skills listed in table 6.1, with median effect sizes greater than d = 0.40

Table 6.2
Summary of the cognitive consequences of game playing

Type of game	Type of test	Number of comparisons	Median effect size
First-person shooter	Perceptual attention	18	**1.18**
Brain training	Perceptual attention	5	0.31
Spatial action	Perceptual attention	6	0.25
Spatial puzzle	Perceptual attention	5	0.15
Real-time strategy	Perceptual attention	9	–0.10
Spatial puzzle	2-D mental rotation (*Tetris* shapes)	6	**0.82**
Spatial puzzle	2-D mental rotation (card rotation)	5	0.38
Real-time strategy	2-D mental rotation (2-D shapes)	3	–0.03
First-person shooter	2-D mental rotation (*Tetris* shapes)	1	–0.47
Spatial puzzle	3-D mental rotation	4	0.20
Other spatial	3-D mental rotation	4	**0.80**
Spatial puzzle	Spatial cognition	15	0.04
Other spatial	Spatial cognition	13	0.27
Real-time strategy	Spatial cognition	3	–0.42
Brain training	Spatial cognition	8	0.03
First-person shooter	Executive function	2	0.38
Brain training	Executive function	2	**1.04**
Spatial & massively multiple online	Executive function	3	0.35
Real-time strategy	Executive function	11	0.18
Miscellaneous	Reasoning tasks	10	–0.11
Miscellaneous	Motor tasks	9	0.18
Miscellaneous	Memory tasks	10	0.05

Note: Boldface indicates that the effect size is greater than d = 0.40.

highlighted in boldface. Of the twenty-two comparisons listed in table 6.2, only four meet this criterion, which is considered the minimum effect size for educational relevance (Hattie, 2009). Concerning perceptual attention skills, playing first-person shooter games tended to cause major improvements in perceptual skills ($d = 1.18$ across eighteen comparisons), although playing brain-training, spatial action, spatial puzzle, and real-time strategy games did not. Concerning mental rotation, playing *Tetris* appeared to cause a major improvement in 2-D mental rotation with *Tetris* and non-*Tetris* shapes ($d = 0.82$ across six comparisons), but not with other forms of mental rotation. In contrast, real-real strategy games and first first-person shooter games did not produce major improvements in 2-D mental rotation skills.

Concerning other spatial cognition skills, none of the game categories produced major improvements, including puzzle games, real-time strategy games, and brain-training games. Preliminary research with other games that use 3-D environments suggest that certain games, like first-person shooters, may improve performance on 3-D mental rotation tasks ($d = 0.80$ based on four comparisons), although we need a larger database focused on a specific game genre. The 2-D game *Tetris* did not appear to improve 3-D mental rotation. Concerning executive function skills, there was a hint of preliminary support for playing brain-training games ($d = 1.04$ based on two comparisons), even though interpretation of this effect must be tempered by the fact that it is based on a single experiment. No other game types, including real-time strategy games, produced strong effects on executive function, although further research is needed. There was no strong evidence that game playing improves reasoning, motor, or memory skills.

The following sections examine these results in more detail, although an experiment-by-experiment description is beyond the space limitations of this review, particularly for null effects. For this reason, tables 6.3 through 6.11 provide details about the publication source, the game used for the game and comparison groups, the amount of time spent on game playing, the age category of the players, the test used to measure the cognitive skill, and the effect size.

Effects of Playing First-Person Shooter Games on Perceptual Attention Skills

Table 6.3 summarizes the effects of game playing on perceptual attention skills and is divided into five sections based on type of game—first-person shooter, brain-training, spatial action, spatial puzzle, and real-time strategy games. The first and most important section asks, Does playing a first-person shooter game (such as *Unreal Tournament*) cause improvements in players' perceptual attention skills?

To answer this question, the first step is to clearly define what we mean by a first-person shooter game and what do we mean by perceptual attention skills. A first-person shooter game is a game that is played from a first-person perspective where the objective of the game is to survive in various environments that can range from being on a foreign planet to being a soldier in a modern warfare scenario. During the story mode of first-person shooter games, players must navigate 3-D environment levels while killing enemy nonplayer characters (NPCs) controlled by the game and avoiding being killed themselves. The players accumulate different types of guns, which have different firing distances, firing rates, damage strengths, and ammunition amounts. First-person shooter games also offer other modes of game play in which the player can battle against other players, enemy nonplayer characters, or a mixture of the two. These games often allow players to alter the rules for the level such as number of enemy units, how difficult it is to kill enemies, the goal of the particular level (e.g., kills or capturing objects or locations), whether the match ends after a certain number of points is reached or after a certain period of time, and the types of guns that can be used. For example, in the studies reported by Green and Bavelier (2006a), the game group played *Unreal Tournament 2004* on death-match mode (the highest number of kills wins) for twenty-minute play sessions on the game level known as Rankin (a multilevel indoor environment). Figure 6.2 shows a sample screenshot from this level of the game.

Perceptual attention skills are measured by visual perceptual tasks, such as those listed under the "test" column in table 6.3. Achtman, Green, and Bavelier (2008) identified five visual perceptual skill areas most closely related to video game playing: visuospatial attention, dynamics of visual attention, number of objects that can be attended to, spatial resolution of visual processing, and temporal resolution of visual processing. First,

Table 6.3
Effects of game playing on perceptual attention skills

Source	Game	Time	Control	Test	Age (subgroup)	Effect size
First-person shooter games						
Green & Bavelier (2003, expt. 5)	*Medal of Honor*	10 hrs	*Tetris*	Enumeration	College	1.29
Green & Bavelier (2003, expt. 5)	*Medal of Honor*	10 hrs	*Tetris*	Useful field of view (UFOV)	College	2.61
Green & Bavelier (2006b, expt. 2)	*Medal of Honor*	10 hrs	*Tetris*	Enumeration	College	2.03
Green & Bavelier (2006b, expt. 5)	*Unreal Tournament*	30 hrs	*Tetris*	Multiple object tracking (MOT)	College	1.34
Green & Bavelier (2006a, expt. 3)	*Unreal Tournament*	30 hrs	*Tetris*	Useful field of view	College	0.95
Green & Bavelier (2007)	*Unreal Tournament*	30 hrs	*Tetris*	Spatial resolution	College	1.24
Feng et al. (2007)	*Medal of Honor*	10 hrs	*Balance*	Useful field of view	College (female)	1.85
Feng et al. (2007)	*Medal of Honor*	10 hrs	*Balance*	Useful field of view	College (male)	1.00
Boot, Kramer, Simons, Fabian, & Gratton (2008)	*Medal of Honor*	21.5 hrs	No game	Useful field of view	College	–0.61
Boot et al. (2008)	*Medal of Honor*	21.5 hrs	No game	Attentional blink	College	0.27
Boot et al. (2008)	*Medal of Honor*	21.5 hrs	No game	Multiple object tracking	College	0.26
Boot et al. (2008)	*Medal of Honor*	21.5 hrs	No game	Enumeration	College	0.32
Boot et al. (2008)	*Medal of Honor*	21.5 hrs	No game	Visual short-term memory	College	0.12
Li et al. (2009)	*Unreal Tournament* and *Call of Duty 2*	50 hrs	*The Sims*	Contrast sensitivity	Adults	1.43
Li et al. (2009)	*Unreal Tournament* and *Call of Duty 2*	50 hrs	*The Sims*	Critical period	Adults	1.13

Nelson & Strachan (2009)	*Unreal Tournament*	1 hr	*Portal*	Perceptual speed (RT)	College	1.93
Nelson & Strachan (2009)	*Unreal Tournament*	1 hr	*Portal*	Perceptual speed (RT)	College	1.97
Wu et al. (2012)	*Medal of Honor*	10 hrs	*Balance*	Useful field of view	College	0.39
Median						1.18
Brain-training games						
Lorant-Royer, Munch, Mescle, & Lieury (2010)	*Dr. Kawashima's Brain Training*	8 hrs, 15 mins	No game	Cancellation	Elementary	–0.38
Lorant-Royer et al. (2010)	*Dr. Kawashima's Brain Training*	8 hrs, 15 mins	No game	Symbols	Elementary	0.31
Nouchi et al. (2012)	*Brain Age*	5 hrs	*Tetris*	Cancellation	Older adults	0.30
Nouchi et al. (2012)	*Brain Age*	5 hrs	*Tetris*	Digit symbol coding	Older adults	1.22
Nouchi et al. (2012)	*Brain Age*	5 hrs	*Tetris*	Symbol search	Older adults	0.84
Median						0.31
Spatial action games						
Orosy-Fildes & Allan (1989)	*Centipede*	15 mins	No game	Reaction time	College	1.18
Larose, Gagnon, Ferland, & Pepin (1989)	*Super Breakout*	12 hrs	No game	Visual scanning	Elementary	0.51
Larose et al. (1989)	*Super Breakout*	12 hrs	No game	Visual tracking	Elementary	0.30
Lorant-Royer et al. (2010)	*New Super Mario Brothers*	8 hrs, 15 mins	No game	Cancellation	Elementary	0.19
Lorant-Royer et al. (2010)	*New Super Mario Brothers*	8 hrs, 15 mins	No game	Symbols	Elementary	0.01
Whitlock et al. (2012)	*World of Warcraft*	14 hrs	No game	Perceptual speed	Older adults	–0.02
Median						0.25

Table 6.3 (continued)

Source	Game	Time	Control	Test	Age (subgroup)	Effect size
Spatial puzzle games						
Boot et al. (2008)	*Tetris*	21.5 hrs	No game	Useful field of view	College	–0.47
Boot et al. (2008)	*Tetris*	21.5 hrs	No game	Attentional blink	College	0.15
Boot et al. (2008)	*Tetris*	21.5 hrs	No game	Multiple object tracking	College	0.38
Boot et al. (2008)	*Tetris*	21.5 hrs	No game	Enumeration	College	–0.09
Boot et al. (2008)	*Tetris*	21.5 hrs	No game	Visual short-term memory	College	0.16
Median						0.15
Real-time strategy games						
Basak et al. (2008)	*Rise of Nations*	23.5 hrs	No game	Useful field of view	Older adults	0.11
Basak et al. (2008)	*Rise of Nations*	23.5 hrs	No game	Attentional blink	Older adults	–0.17
Basak et al. (2008)	*Rise of Nations*	23.5 hrs	No game	Enumeration	Older adults	–0.10
Boot et al. (2008)	*Rise of Nations*	21.5 hrs	No game	Useful field of view	College	–0.49
Boot et al. (2008)	*Rise of Nations*	21.5 hrs	No game	Attentional blink	College	0.27
Boot et al. (2008)	*Rise of Nations*	21.5 hrs	No game	Multiple object tracking	College	0.31
Boot et al. (2008)	*Rise of Nations*	21.5 hrs	No game	Enumeration	College	0.02
Boot et al. (2008)	*Rise of Nations*	21.5 hrs	No game	Visual short-term memory	College	–0.17
Glass, Maddox, & Love (2013)	*Starcraft*	40 hrs	*Sims 2*	Visual search	College	–0.13
Median						–0.10

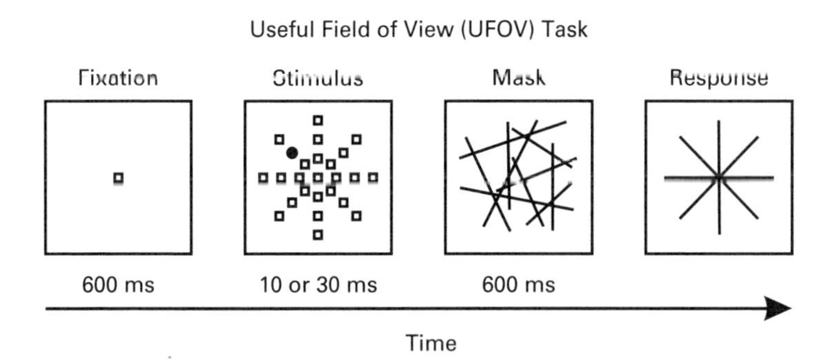

Figure 6.3
The UFOV task used to measure perceptual skill. *Source:* From Feng, Spence, & Pratt (2007)

visuospatial attention is the ability to efficiently distribute attention across a visual field. This skill is typically measured with visual search tasks such as a useful-field-of-view (UFOV) task (Achtman et al., 2008). Nine of the reported comparisons in table 6.3 involve the UFOV task as a measure of perceptual skill. In the UFOV task used by Feng, Spence, and Pratt (2007), for example, the participants first see a fixation square at the center of the screen followed 600 milliseconds later by an array of twenty-four squares, which is presented for ten to thirty milliseconds, as summarized in figure 6.3. During the stimulus presentation trial, one of the squares is randomly selected as the target and filled with a dark-gray color. The target screen is then replaced with a mask for 600 milliseconds followed by a response screen in which the participants have to indicate which direction the target appeared out of the eight possible choices. Better performance on this task is indicated by higher accuracy, especially for targets that are positioned at an eccentricity displaced further from the center.

The attentional blink paradigm has been used to measure the dynamics of visual attention (Achtman et al., 2008). Table 6.3 includes four comparisons using versions of attentional blink tasks. In an attentional blink task, the participants must try to attend to two events within a string of constant stimuli (e.g., two letters in a stream of numbers or the appearance of an X after a particular letter printed in a particular color), as demonstrated by figure 6.4. The participants are not always able to report the second stimulus when it occurs too soon after the presentation of the first, and this

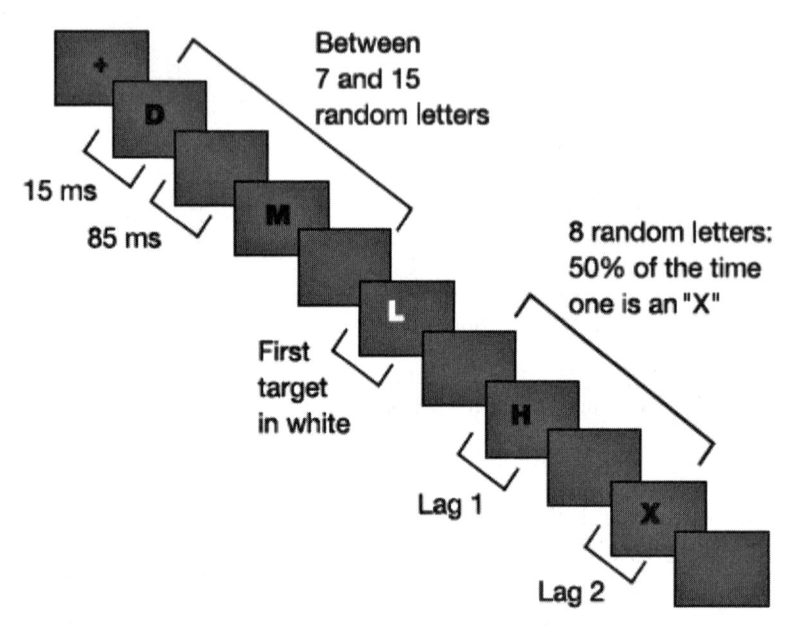

Figure 6.4
The attentional blink task used to measure perceptual skill

period is referred to as their *blink* (Anderson & Bavelier, 2011). The task is designed so that the second target occurs at varying time intervals after the first object, therefore measuring how quickly the participants can recover their attentional resources after attending to the first target (i.e., measuring how long their blink is). Shorter blinks mean that the participants can quickly recover their attentional resources and process the second stimulus.

The third skill that Achtman et al. (2008) discuss as being improved by video games is the ability to attend to multiple objects at one time. In research on video games, two tasks that have been used to examine this skill are the multiple object tracking (MOT) task and the enumeration task. Ten of the comparisons discussed in table 6.3 include data from MOT and enumeration tasks. During the MOT task, the subject is first presented with a static display of similar objects and asked to track a set of distractors (Hubert-Wallander, Green, & Bavelier, 2010). Then, during the trials a set of targets move around randomly along with a set of distractors. At the end of each trial, the participants indicate whether a selected object belonged to the target group or the distractor group. For the enumeration task, a display is quickly flashed on the screen, and the participants must report the number of objects that were presented as quickly and accurately as possible (Green & Bavelier, 2006a).

Three of the comparisons in table 6.3 involve measures of the fourth skill, visual processing, which has been measured using text-crowding tasks and contrast sensitivity functions. For text crowding, the participants have to determine the orientation of a letter when it was flanked by distractor letters, in which the experimenters vary the distance between the target and the distractors (Green & Bavelier, 2007). The stimuli were also presented at three varying distances from the fixation point, based on the idea that game players will show greater accuracy for smaller distances (i.e., more crowding) at eccentricities further from the central fixation.

Contrast sensitivity is the ability to detect differences in luminance on a uniform background. This ability can often be compromised, affecting an individual's overall vision. Individuals with high visual sensitivity are able to detect stimuli with low contrast differences. A contrast sensitivity function shows at what contrast level you are able to detect a stimulus depending on the spatial frequency (width of the lines) of a grating pattern.

Finally, perceptual speed (or processing) refers to how quickly an individual can use visually presented information in order to guide their behavior (Anderson & Bavelier, 2011). An extensive review by Dye, Green, and Bavelier (2009) reported that multiple studies with different tasks have shown that video game players have faster reaction times without sacrificing accuracy and that a few studies have also demonstrated that this increase in speed can be trained by action video games. For example, Nelson and Strachan (2009), asked participants to fixate on a point on the screen, and then click on a randomly located target that appears on the screen as quickly and accurately as possible. Measures included both response time and response distance from the target (as an error measure), which allows for examining the trade-off between speed and accuracy.

The first section of table 6.3 provides the strongest and most consistent effect found in this review. Across seventeen of eighteen experimental comparisons, playing a first-person shooter game resulted in greater improvements in perceptual attention skills (such as those described above) than engaging in a control activity, with a median effect size in the large range. The first two lines of the table summarize the results from Green and Bavelier's (2003) seminal paper examing at the effects of game play on visual selective attention, in which playing ten hours of the first-person shooter game *Medal of Honor* resulted in better improvement on the useful-field-of-view and enumeration tasks than a control group that played ten hours of the puzzle game *Tetris*.

The next two lines in the table show the results of two follow-up studies that reported sufficient information to compute effect sizes (Green & Bavelier, 2006b). Similar to the 2003 article, the first study also produced strong effects on an enumeration task for ten hours of playing the first-person shooter game *Medal of Honor* as compared to a *Tetris*-playing control group. In another study, playing the first-person shooter game *Unreal Tournament* for thirty hours produced strong effects for the multiple-object-tracking task as compared to a *Tetris*-playing control group. The next line in the table summarizes the findings that thirty hours of *Unreal Tournament* strongly improved performance on four different versions of the useful-field-of-view task, which we averaged, including with and without distractors as well as with and without an initial focus on the center of the screen (Green & Bavelier, 2006a). Green and Bavelier (2007) reported similar outcomes, in which thirty hours of playing *Unreal Tournament* resulted in greater improvement in spatial resolution as compared to playing *Tetris*.

The next two lines show results from Feng et al. (2007) in which both men and women who played the 3-D, first-person shooter *Medal of Honor: Pacific Assault* for ten hours outperformed a control group on a useful-field-of-view task (and on a 3-D mental rotation task, which is reported in table 6.3). Similar increases between the two tasks led Feng et al. (2007.) to argue that performance on mental rotation tasks is affected by lower-level abilities such as being able to distribute visual attention, which may be trainable using first-person shooter games. As summarized in the next five lines of table 6.3, Boot, Kramer, Simons, Fabian, and Gratton (2008) found no significant benefit of training with the first-person shooter game *Medal of Honor: Allied Assault* for over twenty hours on any of the attention measures in their cognitive battery including—useful field of view, attentional blink, multiple object tracking, enumeration, or visual short-term memory. The participants in both groups were tested three times—pretest, halfway through the game training, and posttest after training completion—which may account for why the control group showed as much improvement as the game group.

The next two lines in table 6.3 include effect sizes from Li et al. (2009), who found that fifty hours of playing the first-person shooter games *Unreal Tournament* and *Call of Duty 2* resulted in improvements in two measures of contrast sensitivity as compared to playing the simulation game *The Sims 2*. The next two lines summarize two studies reported by Nelson and Strachan (2009) in which participants who played an hour of the first-person shooter game *Unreal Tournament* outperformed those who played the puzzle game *Portal* on tests of perceptual speed. In both experiments, the participants were faster but less accurate after playing *Unreal Tournament* and slower but more accurate after playing *Portal*. It should be noted that this study did not necessarily train the participant's perceptual speed but caused a strategy shift during the tasks. Longer training times would be needed to see how a speed-accuracy trade-off might change over time for perceptual speed and first-person shooter playing.

The final study in the first-person shooter game section of table 6.3 comes from Wu et al. (2012), who found that ten hours of playing *Medal of Honor: Pacific Assault* resulted in moderate improvements on the useful-field-of-view task as compared to a control group that played the puzzle game *Balance*. This study is noteworthy for its use of electroencephalography to measure event-related potentials (i.e., changes in the electrical waveform caused by the presentation of a stimulus or event) on the useful-field-of-view task before and after game playing in both groups. Behaviorally, the participants in the action game group showed greater improvement on the use-field-of-view task. For the event-related potential (ERP) data, the results indicated that action game players did not show any differences in the event-related potentials that are used to measure early processing of sensory information such as the deployment of selective attention. The study did find that in high-performing first-person shooter trainees, there were increased amplitudes for both P2 ($d = 0.68$) and P3 ($d = 0.98$) between the action game players and the puzzle game group. These findings suggest that players who showed high improvement were improving top-down influences on attention such as being able to ignore distractors and efficiently allocate their attentional resources. The authors proposed that a longer training regime may be more likely to affect other elements of visual processing.

As shown in this section, the strongest and most consistent effect we found across all cognitive consequences studies is that playing first-person shooter games improves perceptual attention skills. This result, which can be called the *first-person shooter principle*, is most in line with the specific transfer of general skill view in which playing games that require repeated use of a set of perceptual skills results in improvements of those skills as measured in nongame tasks.

It should be noted that all the data for the first-person shooter game principle come from college students and older adults, and rely mainly on *Medal of Honor* and *Unreal Tournament*. Further research is needed to determine what may be causing these changes and whether other games may exist that include similar elements that can tax the visual system. These issues are partially addressed in the other portions of table 6.3, which focus on the effects of playing computer games that are not first-person shooter games.

Effects of Playing Brain-Training Games on Perceptual Attention Skills

The second portion of table 6.3 lists five experimental comparisons between a group that played a brain-training game versus one that played no game (or a puzzle game) on measures of perceptual attention. Promoters frequently claim that playing brain-training games can improve a variety of cognitive skills, but our review of five comparisons based on two publications (Lorant-Royer, Munch, Mescle, & Lieury, 2010; Nouchi et al. 2012) produced only a small median effect size for measures of perceptual attention skill ($d = 0.31$). In two comparisons, Lorant-Royer et al. (2010) found that eight hours of playing the brain-training game known as *Dr. Kawashima's Brain Training* did not greatly improve performance on two measures of perceptual skill: the cancellation task and the symbol task. The cancellation task is used to evaluate the visual speed of data processing, short-term visual memory, and visual attention. The participants were asked to cross out as many target shapes or numbers as possible during a limited amount of time while ignoring distractors. For the symbols task, participants were asked to view two target symbols and then determine whether these symbols appeared in a series of five symbols.

In contrast, Nouchi et al. (2012) found that playing five hours of the brain-training game *Brain Age* increased performance on two measures of perceptual attention skill (i.e., symbol search and digit symbol coding), but not on another (i.e., cancellation task) as compared to a group that played the puzzle game *Tetris*. For digit symbol coding, the participants were given a key that pairs numbers with symbols. They were then asked to draw as many symbols as they could in 120 seconds under the corresponding numbers. In the symbol search task, participants were given a target group of symbols, and then had 120 seconds to search through a group of symbols and indicate whether any of them match the target group.

One issue with drawing comparisons involving brain-training games is that the experiments often differ in terms of which types of subgames the participants are asked to play. In the Lorant-Royer et al. (2010) study, the participants were allowed to choose which subgames they would like to play, whereas Nouchi et al. (2012) gave their participants specific instructions on which subgames they could play.

Effects of Playing Spatial Action Games on Perceptual Attention Skills

The third section of table 6.3 lists six experimental comparisons between a group that played a spatial action game versus a group that played no game on measures of perceptual attention and speed. The first three lines of the third section of table 6.3 focus on the classic arcade games *Centipede* and *Super Breakout*, while the next two lines focus on *New Super Mario Brothers*, a 2-D platform game played on the portable system Nintendo DS Lite, and the final game in this category is the massively multiplayer online role-playing game *World of Warcraft*, which uses a 3-D environment. Overall, this diverse collection of experiments (Larose, Gagnon, Ferland, & Pepin, 1989; Lorant-Royer et al., 2010; Orosy-Fildes & Allan, 1989; Whitlock et al., 2012) produced only a small median effect size ($d = 0.25$).

The first line of the third portion of table 6.3 lists a study by Orosy-Fildes and Allan (1989) showing that just fifteen minutes of playing the arcade game *Centipede*—which requires the player to react quickly to the movement of the enemy centipede as it progresses from the top of the screen to the bottom, where the player's character fires from—resulted in significantly improved performance on a visual reaction time task. The next two lines feature a study by Larose et al. (1989) in which participants played twelve hours of *Super Breakout*—a simple reaction-based game where the player must move a paddle to hit a ball into a wall of bricks. The ball bounces back, and the player must continue to destroy the blocks until all of them have been eliminated or until the paddle at the bottom of the screen misses the ball. Playing *Super Breakout* resulted in small and nonsignificant improvements on visual scanning and tracking tasks as compared to a control group. As part of a larger study, Lorant-Royer et al. (2010) asked students to play eight hours of *New Super Mario Brothers*, which requires them to navigate their character, typically Mario or Luigi, through side-scrolling, 2-D environments that get progressively harder as the player advances through the levels. The player must be able to fight enemies, avoid projectiles and traps, and solve jumping puzzles. Playing the game did not produce significant improvements in perceptual speed and attention, as measured by cancellation and symbols tasks. Finally, Whitlock et al. (2012) did not find any positive effects of playing *World of Warcraft* for fourteen hours on the perceptual speed of older adults as measured by a digit symbol substitution task.

Effects of Playing Spatial Puzzle Games on Perceptual Attention Skills

The only study reported in the fourth section of table 6.3 on spatial puzzle games is that of Boot et al. (2008), which included a comparison of playing the puzzle game *Tetris* for 21.5 hours to a non-game-playing control. Boot et al. (2008) found no evidence that *Tetris* had any effect on any of the visual attention measures used in their cognitive battery. An interesting thing about spatial puzzle games is that they are often used for control conditions for experiments with first-person shooter games. The majority of comparisons in the first section of table 6.3 compare playing *Medal of Honor* or *Unreal Tournament* to playing *Tetris,* and typically show a benefit for the first-person shooter game. It therefore appears that *Tetris* does not tax the perceptual system enough to cause significant cognitive changes in perceptual attention skills.

Effects of Playing Real-Time Strategy Games on Perceptual Attention Skills

The final section of table 6.3 summarizes three studies that examined whether real-time strategy games could affect visual attention. The first three lines include comparisons based on a study by Basak et al. (2008), which looks at the effects of playing the real-time strategy game *Rise of Nations* for 23.5 hours on older adults. In *Rise of Nations*, participants must manage resources in order to maintain and expand their nation. The players must balance sustaining and defending their cities while attacking other cities in order to claim more of the map. Different types of troops that they can deploy can be more or less effective against enemy units depending on the types. Players can also build "wonders" in their cities to gain nonmilitary prowess for their nation. If a player loses all of the cities in their nation they lose the game, whereas ways to win the game include conquering neighboring nations or controling a certain percentage of the map. Unlike first-person shooter games reviewed in the first section of table 6.3, Basak et al. (2008) found no large effects on useful-field-of-view, attentional blink, and enumeration tasks. The next five comparisons, from the Boot et al. (2008) study, also found no significant effect for playing *Rise of Nations* for 21.5 hours on any of their visual attention measures. In addition, Glass, Maddox, and Love (2013) found that playing the real-time strategy game *Starcraft* did not increase performance on a visual search task.

Overall, in accord with the specific transfer of general skills theory of cognitive transfer, the effects of game playing on perceptual skills are strongest when the cognitive processing that is required in the game corresponds with the cognitive processing that is required in the test. When the goal is to improve perceptual attention and perceptual speed, the most effective games are first-person shooter games since this kind of game taxes the visual processing system. Achtman et al. (2008) proposed that strategy games should not be as effective as first-person shooter games for perceptual attention skills because they are not as fast paced and do not tax the visual system.

Effects of Playing Spatial Puzzle Games on 2-D Mental Rotation

Video games have been proposed as possible training tools to increase spatial abilities, which have been linked to success in the STEM disciplines (Wai et al., 2009). Spatial cognition tasks require participants to mentally manipulate an object such as by rotating, transforming, or rearranging the pieces of the objects, although there is not agreement on how to break spatial cognition into specific factors measured by specific tasks (Hegarty & Waller, 2005; Linn & Petersen, 1985; Uttal et al., 2012). In light of the game research literature, we break the domain of spatial cognition tasks into 2-D mental rotation (summarized in table 6.4), 3-D mental rotation (summarized in table 6.5), and other spatial tasks (summarized in table 6.6).

Table 6.4 summarizes whether groups that play computer games show greater improvements on 2-D mental rotation than groups that engage in other activities, and is divided into sections for spatial puzzle games, real-time strategy games, and first-person shooter games. The only section that yielded strong positive effects is the first one, in which playing the spatial

Table 6.4

Effects of game playing on 2-D mental rotation

Source	Game	Time	Control	Test	Age (subgroup)	Effect size
Spatial puzzle games on 2-D *Tetris* shapes and non-*Tetris* blocks						
Okagaki & Frensch (1994, expt. 2)	*Tetris*	6 hrs	No game	2-D *Tetris* shapes	College (female)	1.28
Okagaki & Frensch (1994, expt. 2)	*Tetris*	6 hrs	No game	2-D *Tetris* shapes	College (male)	0.96
Okagaki & Frensch (1994, expt. 1)	*Tetris*	6 hrs	No game	Non-*Tetris* blocks	College (female)	0.68
Okagaki & Frensch (1994, expt. 1)	*Tetris*	6 hrs	No game	Non-*Tetris* blocks	College (male)	1.34
Sims & Mayer (2002)	*Tetris*	12 hrs	No game	2-D *Tetris* shapes	College	0.59
Boot et al. (2008)	*Tetris*	21.5 hrs	No game	2-D *Tetris* shapes	College	0.52
Median						0.82
Spatial puzzle games on 2-D card rotation						
Okagaki & Frensch (1994, expt. 1)	*Tetris*	6 hrs	No game	2-D card rotation	College (female)	0.38
Okagaki & Frensch (1994, expt. 1)	*Tetris*	6 hrs	No game	2-D card rotation	College (male)	0.12
De Lisi & Wolford (2002)	*Tetris*	5.5 hrs	*Carmen Sandiego*	2-D card rotation	Elementary (female)	1.07
De Lisi & Wolford (2002)	*Tetris*	5.5 hrs	*Carmen Sandiego*	2-D card rotation	Elementary (male)	1.09
Sims & Mayer (2002, expt. 2)	*Tetris*	12 hrs	No game	2-D card rotation	Adult (female)	0.16
Median						0.38
Real-time strategy games on 2-D shapes						
Basak et al. (2008)	*Rise of Nations*	23.5 hrs	No game	Alphanumeric (accuracy)	Older adults	0.60
Basak et al. (2008)	*Rise of Nations*	23.5 hrs	No game	Alphanumeric (RT)	Older adults	–0.03
Boot et al. (2008)	*Rise of Nations*	21.5 hrs	No game	2-D *Tetris* shapes (RT)	College	–0.42
Median						–0.03
First-person shooter game on 2-D *Tetris* shapes						
Boot et al. (2008)	*Medal of Honor*	21.5 hrs	No game	2-D *Tetris* shapes (RT)	College	–0.47

puzzle game *Tetris* resulted in improvements in the mental rotation of *Tetris* and *Tetris*-like shapes. In short, the first section of table 6.4 addresses the question, Does playing a spatial puzzle game (such as *Tetris*) cause improvements in players' mental rotation skill? To answer this question, a first step is to clearly define what we mean by a spatial puzzle game and what we mean by mental rotation.

For our purposes, a spatial puzzle game is any type of video game that requires the participant to manipulate an object within space in order to solve problems and accumulate points. The most used spatial puzzle game in the literature is *Tetris*, as exemplified in figure 6.1 and described previously. One possible reason that *Tetris* has received so much interest in the games literature is the proposed connection to mental rotation (Shepard & Metzler, 1971). In mental rotation tasks, participants are asked to determine whether a single object or set of objects (such as block shapes or letters) are the same (e.g., same object rotated to different orientations) or different (e.g., one is mirror reversed from the other as well as rotated or is a completely different object), than a target object, such as shown in Figure 6.5. For example, in the card rotation test (Ekstrom, French, & Harman, 1976), participants are given a 2-D target shape, and must determine whether eight objects are the same or different (i.e., mirror reversed). Participants have three minutes to complete as many of the items as they can with two testing sets. For 2-D rotation tests, many researchers have developed their own tasks using *Tetris*-like shapes, other shapes, or letters.

The first section of table 6.4 summarizes research on the effects of *Tetris* playing on the 2-D mental rotation of *Tetris*-like shapes, thereby ensuring a tighter correspondence between cognitive processes required in the game and cognitive processes required in the test. In six comparisons, across three different publications there was strong support for the *spatial puzzle*

Are the two shapes the same or different?

SAME DIFFERENT

Figure 6.5
A mental rotation task

game principle, yielding a median effect size of $d = 0.82$. In the first four comparisons from Okagaki and Frensch (1994), the treatment group played six hours of *Tetris*, while the control group was asked to refrain from playing video games during the fourteen days between the pretest and the posttest. The tasks used in this experiment dealt with shapes that appeared in the game *Tetris* as well as non-*Tetris* shapes to determine whether playing *Tetris* improved players' more general spatial skills. In the task, the participants had to mentally rotate *Tetris* and non-*Tetris* shapes in order to make same/different judgments. The results showed that both male and female participants significantly improved their performance on the mental rotation task for both familiar *Tetris*-like shapes and more complex, unfamiliar non-*Tetris* shapes. It is important to note that even though the non-*Tetris* shapes were not shapes taken directly from the game, they were very similar block shapes with one or two squares added or moved.

Sims and Mayer (2002) found that practice with twelve hours of *Tetris* showed no significant effect of training beyond the repeated testing benefits experienced by the control group, although the effect size was in the medium range. Further analysis of the rotation times revealed that the game players altered how they rotated some of the *Tetris* shapes such as rotating clockwise for 225° instead of a less efficient counterclockwise rotation. Sims and Mayer (2002) contend that playing Tetris results in specific skill transfer.

Boot et al. (2008) argue for the case of specific transfer with their 2-D *Tetris* shape rotation task. The participants in their study played *Tetris for* 21.5 hours. On the posttest, the *Tetris* group had significantly faster response times for the mental rotation of 2-D *Tetris* shapes over a control group.

The second portion of table 6.4 summarizes data from experimental comparisons between a group that played *Tetris* and one that played no game or a different kind of game on a posttest (or pretest-to-posttest gain) that involves a more distant measure of transfer for 2-D mental rotation (i.e., the card rotation test). Overall, in five experimental comparisons across three different publications, the spatial puzzle group outperformed the control group, yielding a median effect size of $d = 0.38$, which is a small-to-medium effect.

The first experiment examined whether playing *Tetris* would influence performance on the 2-D card rotations task, and was conducted by Okagaki and Frensch (1994). College-age participants played 6 hours of *Tetris* over two weeks. The results showed that playing *Tetris* did not cause a significant increase in performance for either males or females compared to the non-game-playing control group. Sims and Mayer (2002) also found that performance on the card rotation test did not increase for female graduate students who played *Tetris* for 12 hours. In contrast, De Lisi and Wolford (2002) reported large effects with elementary school boys and girls who played 5.5 hours of *Tetris* while the control group played the text-based *Where in the USA Is Carman Sandiego?* Training was especially useful for female participants considering that males significantly outperformed them on the pretest but not the posttest.

Overall, the findings in the first two sections of table 6.4 provide some evidence for the idea that people can improve their 2-D mental rotation skill by playing spatial puzzle games, although the effects are strongest when mental rotation involves *Tetris*-like shapes rather than nonblock shapes.

Effects of Playing Real-Time Strategy Games on 2-D Mental Rotation

The fourth section of table 6.4 summarizes two studies examining the effects of playing the real-time strategy game *Rise of Nations* on the 2-D mental rotation of alphanumeric stimuli (e.g., letters and digits) and *Tetris*-like shapes. According to the specific transfer of general skill theory, we would not expect a strong effect because the cognitive demands of the game do not strongly match the cognitive processes required for the test. Basak et al. (2008) found that after 23.5 hours of game play, the participants did not significantly differ from the control group in their response time measures for rotating alphanumeric stimuli. The game group, however, did significantly improve on accuracy measures, especially for larger angles. This difference also did not show up when the participants were tested after 11 hours of practice; only during the final posttest after 23.5 hours did video game training have a significant effect. Finally, Boot et al. (2008) found no significant improvements on mental rotation for participants who played *Rise of Nations* for 21.5 hours on either accuracy or response time measures. Overall, there is not consistent evidence that playing real-time strategy games can improve performance on a mental rotation task, so these results support the specific transfer of general skill theory as well as the specific transfer theory.

Effects of Playing First-Person Shooter Games on 2-D Mental Rotation

The final section of table 6.4 shows that playing through the story mode of the first-person shooter game *Medal of Honor* had no positive effect on 2-D mental rotation performance (Boot et al., 2008). This finding is also consistent with the specific transfer of general skill theory (as well as the specific transfer theory) because mental rotation is not a key skill tapped in this version of the game.

Effects of Playing Spatial Puzzle or Other Spatial Games on 3-D Mental Rotation

Table 6.5 extends the question raised in table 6.4 by asking whether game playing can improve 3-D mental rotation skill—that is, when the shapes are presented in three dimensions. The tasks include Shepard and Metzler mental rotation tasks (Shepard & Metzler, 1971), in which students compare 3-D block figures; Vandenberg-Kuse mental rotation

Table 6.5
Effects of game playing on 3-D mental rotation

Source	Game	Time	Control	Test	Age (subgroup)	Effect size
Spatial puzzle games						
McClurg & Chaille (1987)	*The Factory*	9 hrs	No game	Shepard MRT	Elementary	0.36
Okagaki & Frensch (1994, expt. 1)	*Tetris*	6 hrs	No game	Cube comparison	College (female)	0.22
Okagaki & Frensch (1994, expt. 1)	*Tetris*	6 hrs	No game	Cube comparison	College (male)	1.00
Terlecki et al. (2008)	*Tetris*	12 hrs	*Solitaire*	VMRT	College	0.18
Median						0.20
Other spatial games						
McClurg & Chaille (1987)	*Steller 7*	9 hrs	No game	Shepard MRT	Elementary	0.40
Feng et al. (2007)	*Medal of Honor*	10 hrs	*Balance*	VMRT	College (female)	1.20
Feng et al. (2007)	*Medal of Honor*	10 hrs	*Balance*	VMRT	College (male)	1.87
Whitlock et al. (2012)	*World of Warcraft*	14 hrs	No game	VMRT	Older adults	0.01
Median						0.80

tasks (VMRT; Vandenberg & Kuse, 1978), in which students identify whether 3-D block shapes are rotated versions of the target shape versus being different shapes; and cube comparison tasks (Ekstrom et al., 1976), in which students compare 3-D cubes that have a letter or digit on each face.

The top section of table 6.5 shows that playing a spatial puzzle game such as *Tetris* did not have a consistently strong effect on 3-D mental rotation performance (d = 0.20 based on four comparisons). Overall, in most comparisons, practice manipulating 2-D shapes in *Tetris* does not appear to transfer to the mental rotation of 3-D shapes, suggesting that cognitive skills can be improved in a very narrow range. For Terlecki et al. (2008), it is important to note that although there were no significant differences on the Vandenberg-Kuse mental rotation task between *Tetris* players and participants who played *Solitaire*, the results with other spatial cognition tasks in their cognitive battery suggest that *Tetris* caused a wider transfer of spatial skills to other tasks. The experiment examined pairing repeated mental rotation testing over twelve weeks with twelve hours of playing with either video game. Pretest-posttest gains on the Guildford-Zimmerman Spatial Visualization Task and the Surface Development Task (SDT) from Ekstrom, French, & Harman (1976) suggests that a more generalized spatial ability was developed through playing *Tetris*. It is unclear, however, whether repeated exposure to the mental rotation test played a significant role in this process. Lastly, given that Okagaki and Frensch (1994) found strong effects for men, more research is needed to determine how to produce positive game effects on 3-D mental rotation for all players.

The second section of table 6.5 reports research on other how other spatial games affect performance on 3-D mental rotation tasks. While *Stellar 7* and *Medal of Honor* both showed improvements (Feng et al., 2007; McClurg & Chaille, 1987), the massively multiplayer online role-playing game *World of Warcraft* did not result in any significant differences (Whitlock et al., 2012). All three of these games use 3-D environments, but that may not guarantee that the participants will improve their spatial cognition skills. Feng et al. (2007) suggest that one possible reason first-person shooter games may influence spatial cognition is that they can train the lower-level abilities, such as the distribution of visual attention, that may affect performance on spatial cognition tasks. In their study, ten hours of playing *Medal of Honor: Pacific Assault* versus the puzzle game *Balance* resulted in higher performance on the Vandenberg-Kuse mental rotation task (as well as the useful-field-of-view task). It is unclear how the demands of different spatial cognition tasks may affect the benefits from playing first-person shooter games. Overall, the bottom section of table 6.5 provides a hint of a positive effect of playing certain spatial games on 3-D mental rotation ($d = 0.80$ over four comparisons), but a larger database is needed in order to draw solid conclusions. In particular, further research is needed to examine what game elements are necessary to foster improvements in 3-D mental rotation performance.

Effects of Playing Spatial Puzzle, Other Spatial, Real-Time Strategy, and Brain-Training Games on Spatial Cognition Skills

Although mental rotation has been tested in many game studies, some studies have included other spatial cognition tasks mainly based on the classic cognitive factor of visualization (Carroll, 1993)—the ability to mentally manipulate mental images (such as those measured by the paper-folding, form-board, or surface-development tests). In addition, some studies examine visual memory—the ability to remember presented visual information (such as the spatial span and spatial two-back tasks); perceptual speed—the ability to make rapid perceptual judgments; and spatial orientation—the ability to take another perspective.

Table 6.6 summarizes research on the issue of whether playing various kinds of games can affect performance on a variety of spatial cognition tasks (as listed under the "test" column of table 6.6). In the interests of conciseness, we can summarize the findings by saying that there is not strong evidence that spatial cognition skills can be improved by playing *Tetris* ($d = 0.04$ based on fifteen comparisons; Boot et al., 2008; Okagaki & Frensch, 1994; Sims & Mayer, 2002; Subrahmanyam & Greenfield, 1994; Terlecki et al., 2008), other spatial games ($d = 0.27$ based on thirteen comparisons; Boot et al., 2008; Dorval & Pepin, 1986; Gagnon, 1985; Lorant-Royer et al., 2010; Whitlock et al., 2012), real-time strategy games ($d = -0.42$ based on three comparisons; Boot et al., 2008), or brain-training games ($d = 0.03$ based on eight comparisons; Lorant-Royer et al., 2010; Owen et al., 2010).

Effects of Playing First-Person Shooter, Brain-Training, and Other Spatial Games on Executive Function Skills

Does playing a first-person shooter game such as *Unreal Tournament* or spatial games such as *Tetris* improve the player's executive function skills, like

Table 6.6
Effects of game playing on spatial cognition tasks

Source	Game	Time	Control	Test	Age (subgroup)	Effect size
Spatial puzzle games						
Subrahmanyam & Greenfield (1994)	*Marble Madness*	2 hrs, 15 mins	No game	Dynamic spatial	Elementary	–0.01
Okagaki & Frensch (1994, expt. 1)	*Tetris*	6 hrs	No game	Form board	College (female)	–0.13
Okagaki & Frensch (1994, expt. 1)	*Tetris*	6 hrs	No game	Form board	College (male)	0.76
Okagaki & Frensch (1994, expt. 1)	*Tetris*	6 hrs	No game	Perceptual speed	College (female)	–0.23
Okagaki & Frensch (1994, expt. 1)	*Tetris*	6 hrs	No game	Perceptual speed	College (male)	–0.10
Okagaki & Frensch (1994, expt. 2)	*Tetris*	6 hrs	No game	Visualization	College (female)	0.80
Okagaki & Frensch (1994, expt. 2)	*Tetris*	6 hrs	No game	Visualization	College (male)	0.52
Sims & Mayer (2002)	*Tetris*	12 hrs	No game	Form board (*Tetris* shapes)	College	0.27
Sims & Mayer (2002)	*Tetris*	12 hrs	No game	Form board (non-*Tetris* shapes)	College	0.09
Sims & Mayer (2002)	*Tetris*	12 hrs	No game	Form board	College	0.14
Sims & Mayer (2002)	*Tetris*	12 hrs	No game	Paper folding	College	0.04
Terlecki et al. (2008)	*Tetris*	12 hrs	*Solitaire*	Surface development	College	0.66

Table 6.6 (continued)

Source	Game	Time	Control	Test	Age (subgroup)	Effect size
Boot et al. (2008)	*Tetris*	21.5 hrs	No game	Spatial two-back (RT)	College	0.04
Boot et al. (2008)	*Tetris*	21.5 hrs	No game	Spatial two-back (accuracy)	College	–0.18
Boot et al. (2008)	*Tetris*	21.5 hrs	No game	Block tapping	College	–0.78
Median						0.04
Other spatial games						
Gagnon (1985)	*Targ* and *Battlezone*	5 hrs	No game	Spatial orientation	College	0.15
Gagnon (1985)	*Targ* and *Battlezone*	5 hrs	No game	Visualization	College	0.34
Gagnon (1985)	*Targ* and *Battlezone*	5 hrs	No game	Visual pursuit	College	0.13
Dorval & Pepin (1986)	*Zaxxon*	40 games	No game	Spatial relations	College	0.62
Boot et al. (2008)	*Medal of Honor*	21.5 hrs	No game	Spatial two-back (RT)	College	–0.05
Boot et al. (2008)	*Medal of Honor*	21.5 hrs	No game	Spatial two-back (Accuracy)	College	0.00
Boot et al. (2008)	*Medal of Honor*	21.5 hrs	No game	Block tapping	College	–0.20
Lorant-Royer et al. (2010)	*New Super Mario Brothers*	8 hrs, 15 mins	No game	Visual span	Elementary	0.73
Lorant-Royer et al. (2010)	*New Super Mario Brothers*	8 hrs, 15 mins	No game	Forward spatial span	Elementary	0.43
Lorant-Royer et al. (2010)	*New Super Mario Brothers*	8 hrs, 15 mins	No game	Backward spatial span	Elementary	–0.28
Lorant-Royer et al. (2010)	*New Super Mario Brothers*	8 hrs, 15 mins	No game	Spatial memory	Elementary	0.61
Whitlock et al. (2012)	*World of Warcraft*	14 hrs	No game	Spatial orientation	Older adults	0.27
Whitlock et al. (2012)	*World of Warcraft*	14 hrs	No game	Paper folding	Older adults	0.40
Median						0.27
Real-time strategy games						
Boot et al. (2008)	*Rise of Nations*	21.5 hrs	No game	Spatial two-back (RT)	College	–0.45
Boot et al. (2008)	*Rise of Nations*	21.5 hrs	No game	Spatial two-back (accuracy)	College	0.00
Boot et al. (2008)	*Rise of Nations*	21.5 hrs	No game	Block tapping	College	–0.42
Median						–0.42

Brain-training games						
Lorant-Royer et al. (2010)	*Dr. Kawashima's Brain Training*	8 hrs, 15 mins	No game	Visual span	Elementary	0.58
Lorant-Royer et al. (2010)	*Dr. Kawashima's Brain Training*	8 hrs, 15 mins	No game	Forward spatial span	Elementary	–0.16
Lorant-Royer et al. (2010)	*Dr. Kawashima's Brain Training*	8 hrs, 15 mins	No game	Backward spatial span	Elementary	–0.48
Lorant-Royer et al. (2010)	*Dr. Kawashima's Brain Training*	8 hrs, 15 mins	No game	Visuospatial memory (bicube)	Elementary	0.34
Owen et al. (2010, condition 1)	*Dr. Kawashima's Brain Training*	24 sessions	*Trivia*	Spatial working memory	Adults	0.04
Owen et al. (2010, condition 1)	*Dr. Kawashima's Brain Training*	24 sessions	*Trivia*	Paired associates learning	Adults	0.01
Owen et al. (2010, condition 2)	*Dr. Kawashima's Brain Training*	24 sessions	*Trivia*	Spatial working memory	Adults	0.02
Owen et al. (2010, condition 2)	*Dr. Kawashima's Brain Training*	24 sessions	*Trivia*	Paired associates learning	Adults	0.11
Median						0.03

the ability to rapidly switch between tasks? Strobach, Frensch, and Schubert (2012) propose that we should examine executive function (or control) skills when looking at video game playing because these skills are involved with managing other cognitive processes. Action video games in particular may be preferable for training these skills because playing action games requires the participants to perform several actions at once within close temporal proximity, manage multiple priorities to achieve goals, and receive and interpret feedback constantly to adjust gameplay (Boot et al., 2008; Strobach et al., 2012). The data summarized in table 6.7 show that this is still a developing field and there is not strong evidence that any specific genre of game consistently results in increases on executive function in cognitive flexibility tasks.

The top portion of table 6.7 summarizes two comparisons involving first-person shooter games. First, in the version of the task-switching task in Boot et al. (2008), participants made judgments on whether a number was even or odd (or high or low) depending on the color of the background. The results showed that 21.5 hours of video game training with the first-person shooter game *Medal of Honor* did not result in significant differences in switching costs compared to the non-game-playing control group. In contrast, Green, Sugarman, Medford, Klobusicky, and Bavelier (2012) found an improvement in task switching in nongame players who played both *Unreal Tournament 2004* and *Call of Duty 2* for 50 hours over 6–14 weeks as compared to a control group that spent the equivalent amount of time playing the simulation game *The Sims 2*. The database concerning the effects of first-person shooter games on executive function is too small and inconsistent, with $d = 0.38$ based on two comparisons.

The second section of table 6.7 examines brain-training games based on a study by Nouchi et al. (2012) comparing training with older adults with *Brain Age* for 5 hours to playing *Tetris* for the same amount of time. To assess executive function, they used the frontal assessment battery at bedside and trail-making test. For the frontal assessment battery there are six subtests, which include measures of inhibitory control, sensitivity to interferences, mental flexibility, conceptualization, and environmental autonomy. In the trail-making test, the participants have to link ascending numbers and letters in order, switching back and forth between the numbers and letters, going 1 to A to 2 to B, and so on. The measure for this task is how long the participants take to complete it. The results showed that

Table 6.7

Effects of game playing on executive function

Source	Game	Time	Control	Test	Age (subgroup)	Effect size
First-person shooter games						
Boot et al. (2008)	*Medal of Honor*	21.5 hrs	No game	Task switching	College	–0.05
Green, Sugarman, Medford, Klobusicky, & Bavelier (2012)	*Unreal Tournament* and *Call of Duty 2*	50 hrs	*Sims 2*	Task switching	College	0.80
Median						0.38
Brain-training games						
Nouchi et al. (2012)	*Brain Age*	5 hrs	*Tetris*	Trail making	Older adults	0.86
Nouchi et al. (2012)	*Brain Age*	5 hrs	*Tetris*	Frontal assessment battery	Older adults	1.22
Median						1.04
Spatial and massively multiple online games						
Goldstein et al. (1997)	*Tetris*	25.5–36.5 hrs	No game	Stroop task	Older adults	0.18
Boot et al. (2008)	*Tetris*	21.5 hrs	No game	Task switching	College	0.35
Whitlock et al. (2012)	*World of Warcraft*	14 hrs	No game	Stroop task	Older adults	0.72
Median						0.35
Real-time strategy games						
Basak et al. (2008)	*Rise of Nations*	23.5 hrs	No game	Task switching	Older adults	1.02
Basak et al. (2008)	*Rise of Nations*	23.5 hrs	No game	Operation span	Older adults	–0.03
Basak et al. (2008)	*Rise of Nations*	23.5 hrs	No game	N-back task	Older adults	0.82
Basak et al. (2008)	*Rise of Nations*	23.5 hrs	No game	Visual short-term memory	Older adults	0.15
Basak et al. (2008)	*Rise of Nations*	23.5 hrs	No game	Stopping task	Older adults	–0.30
Boot et al. (2008)	*Rise of Nations*	21.5 hrs	No game	Task switching	College	–0.09
Glass et al. (2013)	*Starcraft*	40 hrs	*Sims 2*	Stroop task	College	0.53
Glass et al. (2013)	*Starcraft*	40 hrs	*Sims 2*	ANT (drift)	College	0.52
Glass et al. (2013)	*Starcraft*	40 hrs	*Sims 2*	Task switching	College	0.18
Glass et al. (2013)	*Starcraft*	40 hrs	*Sims 2*	Multilocation switching	College	0.20
Glass et al. (2013)	*Starcraft*	40 hrs	*Sims 2*	Ospan	College	0.13
Median						0.18

participants who played *Brain Age* for 5 hours scored significantly higher than those who played *Tetris* on both measures of executive function, yielding a median effect size of $d = 1.04$. This promising preliminary work yielded the only large effect found for executive function measures, but a larger database is needed before drawing substantive conclusions.

In examining other spatial games (shown in the third section of table 6.7), Goldstein et al. (1997) found no significant benefit from playing *Tetris* for older adults on the Stroop task, in which the participants must inhibit an automatic response. However, Whitlock et al. (2012) found that older adults who trained with *World of Warcraft* for 14 hours had significant improvements in their performance on the Stroop task compared to the control condition. Boot et al. (2008) found no significant benefit for training using *Tetris* on a task-switching test. Based on three comparisons, the median effect size is $d = 0.35$, which is too small to be considered practically significant.

Overall, the pattern of results is quite uneven and the dependent measures vary somewhat from study to study, so the database is not clear enough yet to make definitive conclusions about the effects of playing various spatial games on executive function.

Effects of Playing Real-Time Strategy Games on Executive Function Skills

The final section of table 6.7, based on three published studies (Basak et al., 2008; Boot et al., 2008; Glass et al., 2013), shows that there is not consistent evidence to support the claim that playing real-time strategy games promotes executive function ($d = 0.18$ based on eleven comparisons). Boot et al. (2008) found no benefit of playing the real-time strategy game *Rise of Nations* for 23.5 hours on a task-switching test intended to measure executive function. In contrast, Basak et al. (2008) reported that playing *Rise of Nations* for 23.5 hours improved performance on a task-switching test and the N-back task (which is intended to measure memory load capacity). During the N-back task, the participants had to determine whether a letter was the same or different from one presented N trials back (regardless of location, and whether it was upper- or lowercase). Blocks differed in whether the letter appeared one trial just before it (one-back) or the item was presented two trials before (two-back). However, Basak et al. (2008) did not find strong positive effects of playing *Rise of Nations* on either a visual short-term memory task, span task, or stopping task (all intended to measure the capacity of perceptual processing and attention). Basak et al. (2008) argued that the reason that *Rise of Nations* may be effective at training task switching skills is because strategy games require players to maintain multiple sub-goals at a time (managing armies, gathering resources, expanding territory, etc.), and therefore have to constantly shift cognitive resources to juggle multiple items at a time. The pattern of results from Basak et al. (2008) suggest that the game-playing participants are improving skills that involve managing items in working memory, and switching back and forth between task demands, not working memory capacity.

This claim is further supported by the work of Glass et al., (2013) with the real-time strategy game *Starcraft*. Their study manipulated the cognitive demands of the game, either making the participants manage one base while fighting against one enemy camp, or having them switch between managing two bases while defending against two enemy camps. The study found that their measures for cognitive flexibility showed greater improvement when the participants were required to rapidly switch between multiple sources of action and information, although *Starcraft* game play in general displayed increased performance compared to *Sims 2*. Similar to Basak et al. (2008), there was no evidence for improvements in other tasks unrelated to the cognitive demands of the game. Thus, while this experiment did not find any benefit for visual attention skills because it was not as fast paced as a first-person shooter game, it did tax the participants to keep track of multiple tasks with varying demands at the same time. This finding argues for that idea that having a game that places demands on certain skills may help to increase those skills, which is the theme of the specific transfer of general skill theory.

Effects of Game Playing on Reasoning, Motor, and Memory Skills

Although the preponderance of cognitive consequences research has focused on perceptual attention, spatial cognition, and executive function skills, you might also be curious about whether game playing can improve your ability to reason, to perform eye-hand coordination tasks, and memory. As summarized in tables 6.8, 6.9, and 6.10, the current state of research does not show strong positive effects of game playing on reasoning, motor, or memory skills, respectively.

In table 6.8, across ten comparisons from four publications (Basak et al., 2008; Boot et al., 2008; Owen et al., 2010; Whitlock et al., 2012), the median

Table 6.8
Effects of game playing on reasoning tasks

Source	Game	Time	Control	Test	Age (subgroup)	Effect size
Basak et al. (2008)	*Rise of Nations*	23.5 hrs	No game	Raven's matrices	Older adults	0.55
Boot et al. (2008)	*Medal of Honor*	21.5 hrs	No game	Tower of London	College	0.03
Boot et al. (2008)	*Medal of Honor*	21.5 hrs	No game	Raven's matrices	College	–0.38
Boot et al. (2008)	*Rise of Nations*	21.5 hrs	No game	Tower of London	College	–0.35
Boot et al. (2008)	*Rise of Nations*	21.5 hrs	No game	Raven's matrices	College	–0.25
Boot et al. (2008)	*Tetris*	21.5 hrs	No game	Tower of London	College	0.08
Boot et al. (2008)	*Tetris*	21.5 hrs	No game	Raven's matrices	College	–0.43
Owen et al. (2010, condition 1)	*Brain Training*	24 sessions	*Trivia*	Reasoning	Adults	0.17
Owen et al. (2010, condition 2)	*Brain Training*	24 sessions	*Trivia*	Reasoning	Adults	0.22
Whitlock et al. (2012)	*World of Warcraft*	14 hrs	No game	Progressive matrices	Older adults	–0.52
Median						–0.11

effect size on reasoning skills was negative (d = –0.11). In table 6.9, across nine comparisons from two publications (Gagnon, 1985; Lorant-Royer et al., 2010), the median effect size on motor skills was negligible (d = 0.18). In table 6.10, across ten comparisons from six publications (Boot et al., 2008; Glass et al., 2013; Goldstein et al., 1997; Nouchi et al., 2012; Owen et al., 2010; Whitlock et al., 2012), the median effect size on memory skills was negligible (d = 0.05).

These findings are based on a variety of games, so more research may be needed to determine whether certain kinds of games produce positive effects. Nevertheless, this review shows that the high level of positive evidence for the effects of first-person shooter games on perceptual attention skills is not matched by the current state of cognitive consequences research on using games to improve reasoning, motor, and memory skills.

Table 6.9
Effects of game playing on motor tasks

Source	Game	Time	Control	Test	Age (subgroup)	Effect size
Gagnon (1985)	*Targ* and *Battlezone*	5 hrs	No game	Eye-hand coordination	College	0.15
Lorant-Royer et al. (2010)	*New Super Mario Brothers*	8 hrs, 15 mins	No game	Motor dexterity (right hand)	Elementary	0.36
Lorant-Royer et al. (2010)	*New Super Mario Brothers*	8 hrs, 15 mins	No game	Motor dexterity (left hand)	Elementary	–0.3
Lorant-Royer et al. (2010)	*New Super Mario Brothers*	8 hrs, 15 mins	No game	Motor dexterity (both hands)	Elementary	0.11
Lorant-Royer et al. (2010)	*New Super Mario Brothers*	8 hrs, 15 mins	No game	Motor dexterity assembly	Elementary	0.18
Lorant-Royer et al. (2010)	*Dr. Kawashima's Brain Training*	8 hrs, 15 mins	No game	Motor dexterity (right hand)	Elementary	0.75
Lorant-Royer et al. (2010)	*Dr. Kawashima's Brain Training*	8 hrs, 15 mins	No game	Motor dexterity (left hand)	Elementary	0.11
Lorant-Royer et al. (2010)	*Dr. Kawashima's Brain Training*	8 hrs, 15 mins	No game	Motor dexterity (both hands)	Elementary	0.33
Lorant-Royer et al. (2010)	*Dr. Kawashima's Brain Training*	8 hrs, 15 mins	No game	Motor dexterity assembly	Elementary	0.18
Median						0.18

Table 6.10
Effects of game playing on memory tasks

Source	Game	Time	Control	Test	Age (subgroup)	Effect size
Goldstein et al. (1997)	*Tetris*	25.5–36.5 hrs	No game	Sternberg task	Older adults	1.19
Boot et al. (2008)	*Medal of Honor*	21.5 hrs	No game	Working memory operation span	College	0.05
Boot et al. (2008)	*Rise of Nations*	21.5 hrs	No game	Working memory operation span	College	–0.39
Boot et al. (2008)	*Tetris*	21.5 hrs	No game	Working memory operation span	College	–0.44
Owen et al. (2010, condition 1)	*Brain Training*	24.4 sessions	*Trivia*	Verbal short-term memory	Adults	0.05
Owen et al. (2010, condition 2)	*Brain Training*	24.4 sessions	*Trivia*	Verbal short-term memory	Adults	0.18
Nouchi et al. (2012)	*Brain Age*	5 hrs	*Tetris*	Forward digit span	Older adults	0.07
Nouchi et al. (2012)	*Brain Age*	5 hrs	*Tetris*	Backward digit span	Older adults	0.04
Whitlock et al. (2012)	*World of Warcraft*	14 hrs	No game	Everyday cognition battery: Memory	Older adults	–0.35
Glass et al. (2013)	*Starcraft*	40 hrs	*Sims 2*	Digit span	College	–0.22
Median						0.05

Effects of Game Playing on Other Cognitive Tasks

In an effort to look at relatively new and unexplored tasks, table 6.11 shows seven comparisons for cognitive tasks that did not fit under our general categories based on six publications (Bartlett, Vowels, Shanteau, Crow, & Miller, 2009; Glass et al., 2013; Green, Pouget, & Bavelier, 2010; Masson, Bub, & Lalonde, 2011; Miller & Robertson, 2010; Nouchi et al., 2012). These assorted tasks included physics motion judgments, mental computation, multitasking, making probabilistic inferences depending on varying amounts of auditory and visual stimuli, measures of overall cognitive function, and the ability to filter out superfluous visual information. The wide variety of games and cognitive measures used makes it difficult to find a clear pattern, but once again first-person shooter games tended to produce the strongest effects. This suggests that more research is needed to pinpoint the kinds of skills that are affected by playing first-person shooter games.

In particular, the research on probabilistic inferences by Green et al. (2010) provides another possible reason for why first-person shooter games may be effective at training multiple perceptual attention and lower-level skills. The authors suggest that tasks that require the accumulation of evidence more efficiently to increase performance are more likely to show increases in performance due to video game training than learning paradigms that require learning specific solutions. Future exploration into these areas may help illuminate the mechanisms underlying increases in cognitive performance.

Discussion

Practical Contributions

Perhaps the single strongest finding in the cognitive consequences literature is that playing a first-person shooter game for an extended period causes substantial increases on measures of perceptual attention and speed. We call this the *first-person shooter game principle.*

The second strongest finding in the cognitive consequences literature is that playing a spatial puzzle game for an extended period causes modest increases on measures of mental rotation of 2-D shapes. We call this the *spatial puzzle game principle.* Consistent benefits, however, seem to be limited to close transfer tasks such as training with *Tetris* for mentally rotating

Table 6.11
Effects of game playing on various cognitive tasks

Source	Game	Time	Control	Test	Age (subgroup)	Effect size
Masson et al. (2011)	*Enigmo*	6 hrs	*Railroad Tycoon 3*	Physics test (motion of objects)	Middle school	0.31
Miller & Robertson (2010)	*Dr. Kawashima's Brain Training*	13 hrs	No game	Mental computation	Elementary	0.56
Bartlett et al. (2009)	*Red Alert 2*	18 mins	Internet search	Multitasking	College	0.28
Green et al. (2010, expt. 3)	*Unreal Tournament* and *Call of Duty 2*	50 hrs	*Sims 2*	Probabilistic inference (visual)	College	1.29
Green et al. (2010, expt. 3)	*Unreal Tournament* and *Call of Duty 2*	50 hrs	*Sims 2*	Probabilistic inference (auditory)	College	0.97
Nouchi et al. (2012)	*Brain Age*	5 hrs	*Tetris*	Global cognitive status	Older adults	0.05
Glass et al. (2013)	*Starcraft*	40 hrs	*Sims 2*	Information filter	College	–0.20

Tetris-like shapes. The work of Terlecki et al. (2008) suggests that more general visualization skills can be trained when the skill is decontextualized such as having the participant use the skill outside the game as well with repeated testing.

The major practical contribution of research on cognitive consequences is that there is strong support that people can improve their perceptual attention skills by playing action computer games, particularly first-person shooter games such as *Unreal Tournament*. Thus, it is reasonable to ask students to play action computer games that have been shown to be effective when the goal of instruction is to improve perceptual attention skills. There is still no clear story on whether action games can be effective at consistently training spatial cognition skills including mental rotation, although the work of Feng et al. (2007) and Sanchez (2012) suggest this is a research area worth further investigation.

Lastly, the current state of the cognitive consequences literature does not yet encourage the idea that game playing can improve general higher-order cognitive skills such as reasoning and memory. It is unrealistic to assume that game playing causes useful cognitive changes in general, because most games did not produce useful effects on most cognitive measures.

Theoretical Contributions

The major theoretical contribution of cognitive consequences research is that computer games are most effective when there is a close alignment between the specific cognitive processing that is required within the game and that the cognitive processing specified as the learning objective. In short, games are most effective when the cognitive skills performed repeatedly with feedback in the game are similar to the cognitive skills that are tested as learning outcomes. A worthwhile research strategy is to pinpoint cognitive skills that are needed for academic success (such as in STEM fields), conduct cognitive task analyses to isolate the cognitive skills that players practice when playing various computer games, and find matches between cognitive skills required in a game and those required to support academic learning.

This review contributes to our understanding of the nature of transfer—that is, the effect of previous learning on new learning or performance. Specific transfer occurs when some element in the previous learning is the same as is required in the new learning (such as learning mental rotation in a game and then learning a scientific topic that requires mental rotation). General transfer occurs when some previous learning provides a general improvement in the learner's mind that enhances new learning of completely different material. As with other research in cognitive science, cognitive consequences research does not support a theory based on general transfer, but does offer some support for a theory based on specific transfer of general cognitive skill—that is, applying cognitive skills learned in games to similar nongame tasks requiring the same skill (Mayer, 2011; Singley & Anderson, 1989). In short, the current state of research on games for learning indicates that the cognitive skills required in the game should correspond to those required for success on academic tasks.

Methodological Contributions

The cognitive consequences approach to game research has been shown to be a scientifically rigorous methodology that is capable of making useful contributions to the field. Choosing an appropriate activity for the control group is essential for maintaining experimental control, such as playing a game that does not require the targeted skill. Conducting a cognitive task analysis of the game is also a useful step in ensuring alignment between the cognitive processing required in the game and cognitive processing required in the assessment of learning outcome.

Future Contributions

Overall, cognitive consequences research offers the strongest evidence for the positive educational value of video games, but mainly when the goal is to improve basic cognitive processing skills such as perceptual attention or spatial processing. Research is needed to pinpoint the best way to help students develop their cognitive processing skills, including the proper amount and spacing of practice, which game genres and formats are most effective, whether training is more effective for different age groups or genders, whether the effects persist over time, the role of motivation, and any unintended negative consequences of playing action games.

One consideration to address is that many of the commercial games reviewed in this paper are classified as violent, such as first-person shooter games. Often games like these receive interest from researchers due to their hypothesized effect on aggression. In a meta-analytic review, Ferguson

(2007) stated that most research on video games and aggression only highlights the negative effects that these games may have while ignoring the positive effects such as increases in visuospatial cognition. His review found that there was only a 2 percent overlap between violent video game variance and aggressive behavior, and that there was a significant publication bias in this area of the literature. Once this bias is corrected for, there was little to no support for the relationship between action games and aggression. Ferguson (2007) argues that most of the research on violent action games that he reviewed showed increases in visuospatial skills while some studies with nonviolent puzzle games demonstrated no general increase in visuospatial skill, therefore supporting the use of action games over traditional puzzle games. When reviewing the action game literature for visuospatial improvements, even after correcting for publication bias there was still a 13 percent overlap in variance between action game playing and visuospatial cognition. This analysis suggests that it is important to weight the positive aspects of game playing against the negative aspects, or seek ways to create educational games that retain the positive features of first-person shooter games without the negative features.

More recently, Green et al. (2010) have also argued that the enhanced connection between neural layers that take in evidence and those layers that integrate evidence over time for making decisions may be one of the reasons behind enhancements due to video game playing. A fledgling trend in this area of research is to examine the neural underpinnings of changes in cognitive processing due to video game playing (Maclin et al., 2011; Wu et al., 2012).

In short, our answer to the question posed by Loftus and Loftus (1983) about the instructional value of game playing is that "something useful" can be learned when people play video games, but in most cases this has not been shown to happen. Overall, the field is ripe for continued research that pinpoints which kinds of games enhance which kinds of cognitive skills for which kinds of learners under which circumstances.

References

Note: *Articles included in the meta-analysis

Achtman, R. L., Green, C. S., & Bavelier, D. (2008). Video games as a tool to train visual skills. *Restorative Neurology and Neuroscience, 26,* 435–446.

Anderson, A. F., & Bavelier, D. (2011). Action game play as a tool to enhance perception, attention, and cognition. In S. Tobias & J. D. Fletcher (Eds.), *Computer Games and Instruction* (pp. 307–329). Charlotte, NC: Information Age Publishing, Inc.

*Bartlett, C. P., Vowels, C. L., Shanteau, J., Crow, J., & Miller, T. (2009). The effect of violent and non-violent computer games on cognitive performance. *Computers in Human Behavior, 25*(1), 96–102.

*Basak, C., Boot, W., Voss, M., & Kramer, A. (2008). Can training in a real-time strategy video game attenuate cognitive decline in older adults. *Psychology and Aging, 23*(4), 765–777.

Boot, W. R., Blakely, D. P., & Simons, D. J. (2011). Do action games improve perception and cognition? *Frontiers in Psychology, 2,* 1–6.

*Boot, W. R., Kramer, A. F., Simons, D. J., Fabian, M., & Gratton, G. (2008). The effects of video game playing on attention, memory, and executive control. *Acta Psychologica, 129,* 387–398.

Carroll, J. B. (1993). *Human cognitive abilities.* New York: Cambridge University Press.

Cohen, J. (1988). *Statistical power analysis for the behavioral sciences* (2nd ed.). Hillsdale, NJ: Erlbaum.

Cohen, J. E., Green, C. S., & Bavelier, D. (2008). Training visual attention with video games: Not all games are created equal. In H. F. O'Neil & R. S. Perez (Eds.), *Computer games and team and individual learning* (pp. 205–228). Oxford: Elsevier Science.

De Lisi, R., & Wolford, J. L. (2002). Improving children's mental rotation accuracy with computer game playing. *Journal of Genetic Psychology, 163*(3), 272–282.

*Dorval, M., & Pepin, M. (1986). Effect of playing a video game on a measure of spatial visualization. *Perceptual and Motor Skills, 62*(1), 159–162.

Dye, M.W.G., Green, C. S., & Bavelier, D. (2009). Increasing speed of processing with action video games. *Current Directions in Psychological Science, 18*(6), 321–326.

Ekstrom, R. B., French, J. W., & Harman, H. H. (1976). *Manual for kit of factor-referenced cognitive tests.* Princeton, NJ: Educational Testing Service.

*Feng, J., Spence, I., & Pratt, J. (2007). Playing an action video game reduces gender differences in spatial cognition. *Psychological Science, 18*(10), 850–855.

Ferguson, C. J. (2007). The good, the bad, and the ugly: A meta-analytic review of positive and negative effects of violent video games. *Psychiatric Quarterly, 78,* 309–316.

*Gagnon, D. (1985). Videogames and spatial skills: An exploratory study. *Educational Communication and Technology, 33*(4), 263–275.

*Glass, B. D., Maddox, W. T., & Love, B. C. (2013). Real-time strategy game training: Emergence of a cognitive flexibility trait. *PLoS ONE, 8*(8), e70350. doi:10.1371/journal.pone.0070350.

*Goldstein, J., Cajko, L., Oosterbroek, M., Michielsen, M., van Houten, O., & Salverda, F. (1997). Video games and the elderly. *Social Behavior and Personality, 25*(4), 345–352.

*Green, C. S., & Bavelier, D. (2003). Action video game modifies visual selective attention. *Nature, 423*, 534–538.

*Green, C. S., & Bavelier, D. (2006a). Effects of action video game playing on the spatial distribution of visuospatial attention. *Journal of Experimental Psychology: Human Perception and Performance, 32*, 1465–1478.

*Green, C. S., & Bavelier, D. (2006b). Enumeration versus multiple object tracking: The case of action video game players. *Cognition, 101*, 217–245.

*Green, C. S., & Bavelier, D. (2007). Action-video-game experience alters the spatial resolution of vision. *Psychological Science, 18*, 88–94.

*Green, C. S., Pouget, A., & Bavelier, D. (2010). Improved probabilistic inference as a general learning mechanism with action video games. *Current Biology, 20*, 1573–1579.

*Green, C. S., Sugarman, M. A., Medford, K., Klobusicky, E., & Bavelier, D. (2012). The effect of action video game experience of task-switching. *Computers in Human Behavior, 28*, 984–994.

Hattie, J. (2009). *Visible learning*. New York: Routledge.

Hegarty, M., & Waller, D. A. (2005). Individual differences in spatial ability. In P. Shah & A. Miyake (Eds.), *The Cambridge handbook of visuospatial thinking* (pp. 121–169). New York: Cambridge University Press.

Hubert-Wallander, B. P., Green, C. S., & Bavelier, D. (2010). Stretching the limits of visual attention: The case of action video games. *Wiley Interdisciplinary Reviews: Cognitive Science, 2*, 222–230.

Kirsh, D., & Maglio, P. (1994). On distinguishing epistemic from pragmatic action. *Cognitive Science, 18*, 513–549.

*Larose, S., Gagnon, S., Ferland, C., & Pepin, M. (1989). Psychology of computers: XIV. Cognitive rehabilitation through computer games. *Perceptual and Motor Skills, 69*, 851–858.

*Li, R., Polat, U., Makous, W., & Bavelier, D. (2009). Enhancing the contrast sensitivity function through action video game training. *Nature Neuroscience, 12*(5), 549–551.

Linn, M. C., & Petersen, A. C. (1985). Emergence and characterization of sex differences in spatial ability: A meta-analysis. *Child Development, 56*, 1479–1498.

Loftus, G. R., & Loftus, E. F. (1983). *Mind at play: The psychology of video games*. New York: Basic Books.

*Lorant-Royer, S., Munch, C., Mescle, H., & Lieury, A. (2010). Kawashima vs "Super Mario"! Should a game be serious in order to stimulate cognitive aptitudes? *European Review of Applied Psychology, 60*(4), 221–232.

Maclin, E. L., Mathewson, K. E., Low, K. A., Boot, W. R., Kramer, A. F., Fabiano, M., et al. (2011). Learning to multitask: Effects of video game practice on electrophysiological indices of attention and resource allocation. *Psychophysiology, 48*, 1171–1183.

*Masson, M.E.J., Bub, D. N., & Lalonde, C. E. (2011). Video-game training and naive reasoning about object motion. *Applied Cognitive Psychology, 25*(1), 166–173.

Mayer, R. E. (2011). *Applying the science of learning*. Upper Saddle River, NJ: Pearson.

Mayer, R. E., & Wittrock, M. C. (2006). Problem solving. In P. A. Alexander & P. Winne (Eds.), *Handbook of educational psychology* (2nd ed., pp. 287–304). Washington, DC: American Psychological Association.

*McClurg, P. A., & Chaille, C. (1987). Computer games: Environments for developing spatial cognition? *Journal of Educational Computing Research, 3*(1), 95–111.

*Miller, D. J., & Robertson, D. P. (2010). Using a games console in the primary classroom: Effects of "brain training" programme on computation and self-esteem. *British Journal of Educational Technology, 41*(2), 242–255.

National Research Council. (2006). *Learning to think spatially*. Washington, DC: National Academy Press.

National Science Foundation. (2010). *Preparing the next generation of STEM innovators: Identifying and developing the nation's capital*. [NSB-10–33.] Arlington, VA: National Research Board.

*Nelson, R. A., & Strachan, I. (2009). Action and puzzle video games prime different speed/accuracy trade-offs. *Perception, 38*(11), 1678–1687.

*Nouchi, R., Yasuyuki, T., Takeuchi, H., Hashizume, H., Akitsuki, Y., Shigemune, Y., et al. (2012). Brain training game improves executive functions and processing speed in the elderly: A randomized controlled trial. *PLoS ONE, 7*(1), e29676. doi:10.1371/journal.pone.0029676.

*Okagaki, L., & Frensch, P. A. (1994). Effects of video game playing on measures of spatial performance: Gender effects in late adolescence. *Journal of Applied Developmental Psychology, 15*, 33–58.

*Orosy-Fildes, C., & Allan, R. W. (1989). Psychology of computer use: XII. Video-game play. Human reaction time to visual stimuli. *Perceptual and Motor Skills, 69*(1), 243–247.

*Owen, A. M., Hampshire, A., Grahn, J. A., Stenton, R., Dajani, S., Burns, A. S., et al. (2010). Putting brain training to the test. *Nature, 465*(7299), 775–778.

Sanchez, C. A. (2012). Enhancing visuospatial performance through video game training to increase learning in visuospatial science domains. *Psychonomic Bulletin and Review, 19,* 58–65.

Shepard, R. N., & Metzler, J. (1971). Mental rotation of three-dimensional objects. *Science, 171,* 701–703.

*Sims, V. K., & Mayer, R. E. (2002). Domain specificity of spatial expertise: The case of video game players. *Applied Cognitive Psychology, 16,* 97–115.

Singley, M. K., & Anderson, J. R. (1989). *The transfer of cognitive skill.* Cambridge, MA: Harvard University Press.

Spence, I., & Feng, J. (2010). Video games and spatial cognition. *Review of General Psychology, 14,* 92–104.

Strobach, T., Frensch, P. A., & Schubert, T. (2012). Video game practice optimizes executive control skills in dual-task and task-switching situations. *Acta Psychologica, 140,* 13–24.

*Subrahmanyam, K., & Greenfield, P. M. (1994). Effect of video game practice on spatial skills in girls and boys. *Journal of Applied Developmental Psychology, 15,* 13–32.

*Terlecki, M. S., Newcombe, N. S., & Little, M. (2008). Durable and generalized effects of spatial experience on mental rotation: Gender differences in growth patterns. *Applied Cognitive Psychology, 22,* 996–1013.

Uttal, D. H., & Cohen, C. A. (2012). Spatial thinking and STEM education: When, why, and how? In B. Ross (Ed.), *Psychology of learning and motivation* (Vol. 57, pp. 147–181). New York: Academic Press.

Uttal, D. H., Meadow, N. G., Tipton, E., Hand, L. L., Alden, A. R., Warren, C., & Newcombe, N.S. (2012). The malleability of spatial skills: A meta-analysis of training studies. *Psychological Bulletin.* [Advance online publication.] doi: 10.1037/a0028446.

Vandenberg, S. G., & Kuse, A. R. (1978). Mental rotations: A group test of three-dimensional spatial visualization. *Perceptual and Motor Skills, 47,* 599–604.

Wai, J., Lubinski, D., & Benbow, C. P. (2009). Spatial ability for STEM domains: Aligning over fifty years of cumulative psychology knowledge solidifies its importance. *Journal of Educational Psychology, 101*(4), 817–835.

*Whitlock, L. A., McLaughlin, A. C., & Allaire, J. C. (2012). Individual difference in response to cognitive training: Using a multi-modal attentionally demanding game-based intervention for older adults. *Computers in Human Behavior, 28*(4), 1091–1096.

*Wu, S., Cheng, C. K., Feng, J., D'Angelo, L., Alain, C., & Spence, I. (2012). Playing a first-person shooter video game induces neuroplastic change. *Journal of Cognitive Neuroscience, 24*(6), 1286–1293.

7 Media Comparison Approach: Are Games More Effective Than Conventional Media?

Chapter Outline

Summary

The media comparison approach to game research compares the learning outcome performance of students who learned by playing a game versus students who were assigned to learn the same material with conventional media. A review of media comparison game research identified two promising academic domains in which games tend to improve learning more than conventional media based on median effect size: science ($d = 0.69$ based on sixteen comparisons) and second-language learning ($d = 0.96$ based on five comparisons). One unpromising domain was mathematics ($d = 0.03$ based on five comparisons), and two not-yet-promising domains were language arts ($d = 0.32$ based on three comparisons) and social studies ($d = 0.62$ based on three comparisons). In terms of age groups, the median effect size favoring games was 0.34 for elementary school students, 0.58 for secondary school students, and 0.74 for college students. In terms of game type, the median effect size was 0.45 for quiz or puzzle games, 0.62 for simulation games, and 0.72 for adventure games. Finally, in terms of type of control group, the median effect size was 0.12 for computer based, 0.53 for paper based, and 0.63 for classroom based. Overall, the media comparison approach may need to be supplemented by the value-added approach in order to pinpoint which game features are particularly effective.

Introduction

Suppose you wished to help students achieve an educational goal like learning the principles of electric circuits or how to solve math problems. Would it be better to teach this material using an interactive computer game or using conventional school media, such as an in-class presentation or illustrated text lesson? The goal of this review is to address this question. This review summarizes research evidence that pinpoints the conditions under which students learn academic content better from games than from traditional school media, such as for different subject areas, age groups, game types, and kinds of control group.

A central principle of scientific research in education is that "particular research designs and methods are suited for specific kinds of investigations and questions" (Shavelson & Towne, 2002, p. 4). The research question posed in this review is whether students learn academic content better from games or conventional media. A research methodology that appears well suited to address this question is the media comparison approach, in which the academic learning outcomes of students who learn by playing a computer game are compared to the academic learning outcomes of students assigned to learn the same material presented with conventional media, as described in chapter 2. Thus, this review differs from other reviews of game research by focusing solely on studies that use a media comparison approach and provide sufficient information to allow for a computation of effect size using Cohen's (1988) *d* for a measure of academic learning.

Rationale for the Media Comparison Approach to Game Research

According to the *game superiority hypothesis* proposed by some game proponents, learning with games can be more effective than learning with conventional media. In *How Computer Games Help Children Learn*, for example, Shaffer (2006) claims that "the key to solving the current crisis in education will be to use the power of computer and video games to give all children access to experiences and build interest and understanding" (p. 8). Similarly, in *Good Video Games and Good Learning*, Gee (2007) asserts, "Good games are problem-solving spaces that create deep learning, learning that is better than what we often see today in our schools" (p. 10). The theoretical basis for the game superiority hypothesis is that games may be able to motivate students to initiate and maintain a high level of engagement with the material better than traditional classroom or textbook venues (Lepper & Malone, 1987; Malone & Lepper, 1987).

Although strong claims have been made for the potential of games to revolutionalize the way students learn, research evidence is needed to scientifically test those claims. In an early book on the potential of educational games, Abt (1970) offered the observation that "games are effective teaching and training devices" (p. 13), but of course there was not a sufficient body of scientifically rigorous evidence at the time to back up this claim.

Over the past decades, reviews of the educational games for academic learning (Connolly, Boyle, MacArthur, Hainey, & Boyle, 2012; Hannafin

& Vermillion, 2008; Hayes, 2005; Honey & Hilton, 2011; Randel, Morris, Wetzel, & Whitehill, 1992; Tobias, Fletcher, Dai, & Wind, 2011; Vogel et al., 2006; Young et al., 2012) as well as research-based edited books (O'Neil & Perez, 2008; Raessens & Goldstein, 2005; Tobias & Fletcher, 2011; Vorderer & Bryant, 2006; Van Eck, 2010) generally offer cautionary conclusions about the effectiveness of educational games for academic learning. For example, in a recent National Research Council report, *Learning Science through Computer Games and Simulations*, a review of the scientific research base indicated that "there is relatively little research evidence on the effectiveness of simulations and games for learning" (Honey & Hilton, 2011, p. 21). In a recent review of ninety-five game studies, Tobias et al., (2011) concluded, "There is considerably more enthusiasm for describing the affordances of games and their motivating properties than for conducting research to demonstrate that those affordances are used to attain instructional aims.... This would be a good time to shelve the rhetoric about games and divert those energies to conducting needed research" (p. 206).

Media comparison research can reduce the gap between grand claims for the educational benefits of computer games and the cautionary conclusions based on game research evidence. Following Shavelson and Towne's (2002) admonition that educational research should use appropriate methods, this review focuses on the media comparison approach because experimental comparisons have been widely recognized as the best way to address questions about instructional effectiveness (Mosteller & Boruch, 2002; Phye, Robinson, & Levin, 2005; Schneider, Carnoy, Kilpatrick, Schmidt, & Shavelson, 2007). In particular, the most direct way to test causal claims about the game superiority hypothesis is through randomized controlled experiments (Mayer, 2011).

Rational against the Media Comparison Approach to Game Research

Media comparison studies can easily confound media and methods, thereby causing researchers to attribute learning outcome effects to instructional media when they are actually caused by instructional methods. Clark (2001) and Clark, Yates, Early, and Moulton, 2010) have argued against media comparison research on the grounds that instructional media do not cause learning; rather, instructional methods cause learning. For example, O'Neil, Wainess, and Baker (2005) note that "positive findings regarding the educational benefits of games ... can be attributed to instructional design and not to games per se" (p. 462). Thus, media comparison studies may be problematic to the extent that they confound method and media, in which the instructional method used with one medium is different from the instructional method used with another medium. A related methodological problem with media comparison studies is that the quantity and quality of the academic content may differ between the game and control groups, so that one group may receive more relevant academic information than the other. In light of these potential conceptual and methodological problems, this review examines game effects with various kinds of control conditions.

Method

The primary goal of this media comparison review is to survey and summarize the available evidence on whether games are more effective than conventional media. In this review, I analyze all available published papers in which academic learning outcomes are compared between a group that is exposed to academic content by playing an educational game (game group) and a group that is taught the same content through conventional media (conventional group). The procedure for this media comparison review includes evidence collection, evidence selection, evidence coding, and evidence summarizing.

Evidence Collection

During evidence collection, I conducted searches of all major social science online databases including PsychINFO and ERIC, using appropriate keywords such as "computer games," "video games," "serious games," "educational games," "simulation games," and "online games." I also searched the listed references of previous literature reviews (e.g., Connolly et al., 2012; Honey & Hilton, 2011; Randel et al., 1992; Tobias et al., 2011; Vogel et al., 2006; Young et al., 2012) and recent relevant research papers. Additionally, I conducted a search of all "cited by" papers for classic relevant papers and reviews.

As noted in chapter 5, overviews of the research literature on computer games tend to show that the vast majority of the available papers do not report scientifically rigorous research evidence (Clark et al., 2011; O'Neil et al., 2005), although more recent reviews were able to identify a growing number of research reports involving educational games (Tobias et al., 2011; Young et al., 2012). In the present review, I was able to identify a small but growing corpus of media comparison studies papers that meet rigorous scientific criteria.

Evidence Selection

During evidence selection, I eliminated all papers that did not meet the following criteria:

- The independent variable involves a comparison of a game group that is assigned to play a game aimed at teaching academic content versus a conventional group that receives an equivalent learning experience using conventional media aimed at teaching the same content. Both groups must be exposed to the same academic content, with only the instructional medium differing between the groups. This can be seen as addressing the criteria of experimental control and random assignment.
- The dependent measure involves a measure of academic learning outcome performance such as answering questions or solving problems (rather than self-reports or in-game activity). This can be seen as the criterion of appropriate measurement.
- The paper reports a mean (*M*), standard deviation (*SD*), and sample size (*n*) for a measure of learning outcome performance for each group, or contains other statistical information that allows for computing a value of effect size based on Cohen's *d*. This also addresses the criterion of appropriate measurement.
- The instructional content is in an academic area, such as science, mathematics, language arts, or social studies.
- The paper is published in a peer-reviewed research journal or book that is accessible.

Evidence Coding

During evidence coding, I created a database for each experimental comparison between a group that receives a game versus a conventional group that receives the same material delivered via conventional media, such as a classroom tutorial, paper-based lesson, or computer-based presentation. I coded for the type of academic content, type of game, age group of the participants, and type of control group as well as effect size for a measure of transfer. When there were multiple measures of retention or transfer, I recorded them all in the database and used the most representative measure of each. In the interests of simplicity and in concert with previous meta-analyses of instructional effectiveness (Mayer, 2009), I opted to focus primarily on one measure of transfer (if available) for each comparison, because transfer is widely recognized as a paramount educational goal (Anderson et al., 2001).

Evidence Summarization and Interpretation

During evidence summarization, I computed effect size for each experimental comparison using Cohen's (1988) *d*. As noted in chapters 5 and 6, I recognize that some experts recommend computing effect size based on *r* or other measures in the *r* family, such as either omega squared or eta squared (Rosenthal, Rosnow, & Rubin, 2000), on the grounds that correlation-based measures are more general. I prefer to use *d* because all comparisons in this review are between two qualitatively different groups, *d* is somewhat more intuitive to communicate, and *d* can easily be converted into *r*. Where it is not possible to compute *d*, I computed Glass's delta (which is based on using the *SD* of the control group as the denominator). I recognize that Hedges's *g* is sometimes recommended when there are large differences in sample size (Ellis, 2010), but I opted to forego this adjustment in the interests of simplicity and consistency. Finally, some experts call for adjusting dependent measures based on test reliability (Ellis, 2010), but this was not feasible because most studies did not include measures of test reliability.

In each type of game environment, I computed the median effect size. As noted in chapters 5 and 6, I recognize that experts in meta-analysis call for using weighted measures of *d* based on sample size in computing a mean effect size across multiple studies (Ellis, 2010), but I opted to report median effect size in the interests of simplicity and to avoid overweighing any one particular study. I focused on transfer test scores for computing median effect size in light of the central role of transfer for assessing educational effectiveness.

Results

Table 7.1 summarizes the effectiveness of game playing for each of five academic domains: science, second-language learning, mathematics, language arts, and social studies.

Table 7.1
Research using the media comparison approach by content area

Type	Number	Effect size
Content area		
Science	12 out of 16	0.69
Second-language learning	4 out of 5	0.96
Math	3 out of 5	0.03
Language arts	3 out of 3	0.32
Social studies	2 out of 3	0.62

Where Games Work: Two Promising Domains

The primary goal of this review is to identify conditions under which games produce better learning outcomes than conventional media. The top two lines of table 7.1 list science and second-language learning as domains in which medium-to-large effect sizes were found based on a substantial number of studies. Concerning science learning, in twelve out of sixteen comparisons, the game group outperformed the conventional group on a measure of transfer, yielding a median effect size of $d = 0.69$. Concerning second-language learning, in four out of five comparisons, the game group outperformed the conventional group, yielding a median effect size of $d = 0.96$.

Table 7.2 summarizes each of sixteen media comparison studies involving science learning. As shown in the first line of table 7.2, in a study by Ricci, Salas, and Cannon-Bowers (1996), adult military trainees learned basic information about chemical, biological, and radiological defense from playing a quiz game with eighty-eight questions or reading a pocket handbook. The game group outperformed the handbook group on a delayed retention test ($d = 0.78$). A delayed retention test may tap learner understanding similar to a transfer test (Katona, 1940).

As shown in the second line of table 7.2, Parchman, Ellis, Christinaz, and Vogel (2000) asked U.S. Navy electronics trainees to take a four-day course in electronics that was delivered as an interactive computer game (game group) or as a computer-based tutorial (conventional group). On a subsequent transfer test involving principle application, the conventional group strongly outperformed the game group ($d = -0.79$ favoring the conventional group).

Table 7.2
Research using the media comparison approach in science

Source	Game version	Control	Topic	Learners	Effect size
Ricci et al. (1996)	*Quiz Game*	Pocket handbook	Military	Adults	0.78
Parchman et al. (2000)	Computer adventure game	E-tutorial	Electronics	Adults	–0.79
Moreno et al. (2001, expt. 1)	*Design-a-Plant* (adventure game)	E-tutorial	Botany	College	1.03
Moreno et al. (2001, expt. 2)	*Design-a-Plant* (adventure game)	E-tutorial	Botany	High school	0.97
Moreno et al. (2001, expt. 3)	*Design-a-Plant* (adventure game)	E-tutorial	Botany	College	0.76
Swaak et al. (2004)	Physics-of-motion simulation game	Hypertext	Physics	Secondary	–0.43
Evans et al. (2008)	Stoichiometry *Virtual Laboratory* course	Text guide	Chemistry	College	0.42
Hickey et al. (2009, expt. 1)	*Quest Atlantis* (adventure game)	Text lesson	Marine life	Elementary	0.14
Hickey et al. (2009, expt. 2)	*Quest Atlantis* with feedback (adventure game)	Text lesson	Marine life	Elementary	0.97
Barab et al. (2009)	*Quest Atlantis* (adventure game)	E-book	Marine life	College	1.55
Anderson & Barnett (2011)	*Supercharged!* (adventure game)	Guided inquiry	Electromagnetism	College	0.72
Wrzesien & Raya (2010)	Interactive adventure game	Lecture	Mediterranean Sea	Elementary	0.24
Brom et al. (2011)	*Orbis Pictus Bestialis* (scenario game)	Lecture	Animal learning	Secondary	0.67
Hwang et al. (2012)	Butterfly ecology game	Worksheets and Web search	Butterflies	Elementary	2.36
Adams et al. (2012, expt. 1)	*Crystal Island* (adventure game)	Slideshow	Infectious disease	College	–0.57
Adams et al. (2012, expt. 2)	*Cache 17* (adventure game)	Slideshow	Electric devices	College	–0.31
Median					0.69

Moreno, Mayer, Spires, and Lester (2001) asked some high schoolers (in experiment 2) and college students (in experiments 1 and 3) to play a botany simulation game called *Design-a-Plant*, in which they chose the roots, stem, and leaves for plants to live under various environmental conditions (game group). Other students received the same information in the form of a noninteractive computer-based tutorial (conventional group). On a transfer posttest, the game group outscored the tutorial group across the three experiments (*d*s = 1.03, 0.97, and 0.76, respectively), as summarized in lines three through five in table 7.2.

In Swaak, de Jong, and van Joolingen (2004), students learned about the physics of motion by playing a simulation (game group) or studying a hypertext covering the same content (conventional group). The students given hypertext outperformed those given simulations on posttests involving definitions, making predictions for what-if questions, and explaining answers to what-if questions ($d = -0.43$ overall), as summarized in line six of table 7.2.

As shown in line seven, Evans, Yaron, and Leinhardt (2008) asked incoming college students to solve chemistry problems by interacting with a sixteen-level *Virtual Laboratory* simulation (taking twenty to twenty-five hours) or by reading a text than contained sixteen corresponding lessons (taking twelve to fifteen hours). The simulation group outperformed the text group on a problem-solving transfer test ($d = 0.42$), although the simulation group also spent almost twice as much time learning the material.

As displayed in lines eight and nine, Hickey, Ingram-Goble, and Jameson (2009) reported a study in which sixth graders learned about ecology by playing the *Quest Atlantis* adventure game for fifteen hours (game group) or by receiving equivalent expository instruction based on a textbook (conventional group). On a problem-solving test similar to that in the game, the game group did not show a much larger gain than the conventional group in experiment 1 ($d = 0.14$), but did demonstrate a larger gain in experiment 2 ($d = 0.97$), in which the game had enhanced feedback elements.

Barab et al. (2009) asked pairs of college students to spend ninety minutes playing as an avatar in the *Taiga Park* adventure game (a simulated aquatic habitat within the *Quest Atlantis* collection) or by individually reading an electronic textbook that covered the same material on marine science. On a subsequent transfer problem-solving test, the game group strongly outperformed the conventional group ($d = 1.55$), as summarized in the tenth line of table 7.2.

As depicted in line 11, Anderson and Barnett (2011) asked college students (i.e., preservice elementary school teachers) to learn about electromagnetism in two two-hour sessions by playing a simulation game, *Supercharged!* or by using conventional guided inquiry methods. On a transfer posttest on conceptual understanding of electromagnetism, the game group outperformed the conventional group ($d = 0.72$).

In Wrzesien and Raya (2010), forty-eight Spanish sixth graders learned about the Mediterranean Sea by playing an interactive adventure game (game group) or receiving an in-class lecture (conventional group). On an eleven-item multiple-choice test on key concepts, the game group performed slightly better than the conventional group ($d = 0.24$), as shown in the twelfth line of table 7.2.

In Brom, Preuss, and Klement (2011), a hundred Czech high school students received a classroom presentation on animal learning that was supplemented by playing a game (called *Orbis Pictus Bestialis*) involving training a dog or through an additional classroom presentation. The game group outperformed the conventional group on a delayed open-ended test ($d = 0.67$), as shown in the thirteenth line of table 7.2.

In a study by Hwang, Wu, and Chen (2012), summarized in the fourteenth line of table 7.2, elementary school children in Taiwan received a 120-minute lesson on butterfly ecology, and then spent 150 minutes playing a computer game based on applying the lesson content (game group) or completing worksheets and searching the Internet for key terms (conventional group). On a posttest on butterfly ecology, the game group strongly outperformed the control group ($d = 2.36$).

The last two lines in table 7.2 show a study by Adams, Mayer, MacNamara, Koening and Wainess (2012) in which college students played an adventure game (game group) containing scientific content, or viewed a slideshow that presented the same content using the same wording and graphics as in the game but without a game context (conventional group). The slideshow group outperformed the game group on a transfer test in experiment 1, where the game was *Crystal Island*, which teaches about infectious disease ($d = -0.57$), and in experiment 2, where the game was *Cache 17*, which teaches about electromagnetic devices ($d = -0.31$).

Table 7.3 summarizes five comparisons involving second-language learning, which together yield a median effect size of $d = 0.96$. As shown in the first line, Segers and Verhoeven (2003) taught vocabulary words to Dutch kindergarteners who were non-Dutch-speaking immigrants

Table 7.3
Research using the media comparison approach in second-language learning

Source	Game version	Control	Topic	Learners	Effect size
Language arts: Second-language learning					
Segers & Verhoeven (2003)	Interactive puzzle game	Classroom	Dutch vocabulary	Elementary	0.37
Yip & Kwan (2006)	Web-based puzzle games	Classroom	English vocabulary	College	1.39
Neri et al. (2008)	Language-learning simulation game	Classroom	English language	Elementary	–0.41
Liu & Chu (2008)	Language-learning simulation game	Classroom	English language	Secondary	0.96
Suh et al., (2010)	Multiplayer role-playing game (English-language learning)	Classroom	English language	Elementary	1.85
Median					0.96

by having them play games twice a week for fifteen minutes over fifteen weeks (game group) or receive regular classroom instruction on the same vocabulary words (conventional group). In the game group, students viewed a short story that consisted of narration and six pictures, and then played a quiz game in which a parrot asked various questions and a pirate provided advice. The game group performed better than the conventional group on a posttest involving the words in the lessons given a year later ($d = 0.37$).

In Yip and Kwan (2006), as shown in line two of table 7.3, a hundred first-year college engineering students in Hong Kong learned English-language vocabulary in a three-week program meeting twice a week for fifty minutes per session that involved playing games on two Web sites (game group) or participating in regular activity-based classroom instruction with the same vocabulary words (conventional group). On a related vocabulary test, the game group strongly outperformed the conventional group ($d = 1.39$).

As shown in the third line of table 7.3, Neri, Mich, Gerosa, and Giuliani (2008) taught Italian-speaking elementary school students to pronounce English words with an interactive computer-based simulation game (game group) or in regular classroom activity (conventional group). On a subsequent pronunciation test, the game group scored worse than the conventional group ($d = -0.41$).

In Liu and Chu (2008), sixty-four seventh graders in Taiwan learned to speak and listen to English words by playing an interactive computer-based simulation game (game group) or in regular classroom activity (conventional group). On a subsequent speaking and listening posttest, the game group outscored the conventional group ($d = 0.96$), as summarized in line four of table 7.3.

Finally, the last line in table 7.3 summarizes a study by Suh, Kim, and Kim (2010) in which 220 South Korean elementary school students learned English-language skills by playing a massive multiplayer online role-playing game with their classmates (game group) or through traditional face-to-face instruction (conventional group) in two forty-minute sessions per week over two months. In the game group, students created an avatar and worked to save a village from monsters by successfully completing tasks that involved English-language listening, speaking, reading, and writing, along with immediate feedback. In the face-to-face group, students received multimedia lessons in their classrooms with a researcher serving as the teacher to ensure the same content was covered as in the game. On a posttest involving English reading, writing, speaking, and listening, the game group strongly outperformed the conventional group ($d = 1.85$).

Where Games Do Not Work: One Unpromising Domain

A secondary goal of this review is to identify domains in which game playing has been shown to be no more effective than learning with conventional media. The third line of table 7.1 shows that in three out of five comparisons involving mathematics content, the game group outperformed the conventional group, yielding a median effect size of $d = 0.03$. It appears that games may not be effective in the domain of mathematics, although a larger database would be helpful in gauging the robustness of this preliminary observation.

Table 7.4 summarizes five comparisons involving math learning, including computer science. The first line of table 7.4 displays a study by Din and Calao (2001) in which some kindergarteners played forty academic games on a Sony PlayStation for forty minutes per day for five days a week for eleven weeks and played an addition thirty minutes every night, whereas

Table 7.4
Research using the media comparison approach in math

Source	Game version	Control	Topic	Learners	Effect size
Din & Calao (2001)	PlayStation games	Classroom	Math	Elementary	0.03
Van Eck & Dempsey (2003)	Math puzzle game	E-tutorial	Math	Secondary	–0.29
Papastergiou (2009)	Web-based games	E-lesson	Computer science	Secondary	0.54
Sindre et al. (2009)	*Age of Computers*	Paper-based exercises	Computer science	College	–0.02
Chang et al. (2012)	*Millionaire*	Worksheets	Math	Elementary	0.53
Median					0.03

others received the regular academic program. Pretest-to-posttest gains on math ($d = 0.03$) were similar for both groups (although the game group excelled in reading, as shown in table 7.5).

In Van Eck and Dempsey (2003), middle school students played a math game based on an aunt and uncle's remodeling business, or learned the same material (i.e., on number sense, measurement, and geometry) in a computer-based tutorial. On a subsequent math posttest involving a different game context, the game group performed worse than the group that learned with a computer-based tutorial ($d = -0.29$), as shown in the second line of table 7.4.

Papastergiou (2009) taught eighty-eight Greek high school students about how computer memory works in a computer science course either by asking them to play a game or interact with a computer-based lesson that presented the same material. In the game, students moved through rooms by accessing information and answering questions. In the computer-based lesson, students received the same information in the form of a presentation and answered the same questions in quizzes. On a thirty-item posttest that involved using the material, the game group outscored the conventional group ($d = 0.54$), as shown in the third line of table 7.4.

The fourth line of table 7.4 summarizes a study by Sindre, Natvig, and Jahre (2009) in which Norwegian college students taking an introductory-level computer science course learned programming basics by playing a quiz game called *Age of Computers* for ninety minutes or solving paper-based exercises based on the same problems as in the game. On a thirteen-item test of the material, the groups did not differ ($d = -0.02$).

Finally, as shown in the last row of table 7.4, Chang, Wu, Weng, and Sung (2012) taught fifth graders in Taiwan the steps involved in solving mathematics problems over two eighty-minute sessions. Some students learned by playing a quizlike game, *Millionaire* (game group), whereas others learned the same steps and solved the same problems using paper worksheets (conventional group). On a problem-solving posttest, the game group performed better than the conventional one ($d = 0.53$).

Where Games Have Not Been Tested Enough: Two Not-Yet-Promising Features

Another secondary goal of this review is to identify academic domains that have insufficient evidence to determine whether games are effective. The bottom two rows of table 7.1 list two such domains: language arts and social studies. The fourth line of table 7.1 indicates that in three out of three experimental comparisons involving language arts content, the game group outperformed the conventional group, with a median effect size of $d = 0.32$.

Table 7.5 provides a deeper look at the media comparison approach to language arts by summarizing the three experimental comparisons. Din and Calao (2001) asked some kindergarteners to play forty academic games on a Sony PlayStation for forty minutes per day for five days a week for eleven weeks and play an additional thirty minutes every night, whereas others received the regular academic program, as described in the section

Table 7.5
Research using the media comparison approach in language arts

Source	Game version	Control	Topic	Learners	Effect size
Din & Calao (2001)	PlayStation games	Classroom	Spelling	Elementary	0.18
Din & Calao (2001)	PlayStation games	Classroom	Reading	Elementary	1.27
Segers & Verhoeven (2003)	Interactive puzzle game	Classroom	Dutch vocabulary	Elementary	0.32
Median					0.32

on mathematics learning. Pretest-to-posttest gains on spelling ($d = 0.18$) were similar for both groups, as shown in the first line of table 7.5, but the game group showed a much larger gain than the control group in reading, as shown in the second line ($d = 1.27$).

Segers and Verhoeven (2003) taught vocabulary words to Dutch elementary school students who were native speakers of Dutch by having them play games twice a week for fifteen minutes over fifteen weeks or receive regular classroom instruction on the same vocabulary words, as described in the section in second-language learning. The game group performed better than the conventional group on a posttest involving the words in the lessons given a year later, as depicted in the third line of table 7.5 ($d = 0.32$), and as noted in table 7.3, similar results were obtained for non-native-Dutch speakers.

The fifth line of table 7.1 summarizes three comparisons involving social studies, yielding a median effect size of $d = 0.62$. Table 7.6 provides a summary of each experimental comparison that involved social studies content. The first and second lines of table 7.6 summarize a study by Virvou, Katsionis, and Manos (2005), in which ninety fourth graders in Greece played a computer-based adventure game about geography called *VR-ENGAGE* (game group) or learned the same material through an intelligent tutoring system (control group). In the game, the players had to navigate through passages guarded by dragons, who asked them questions such as, "Ethiopia is in Africa, right?" and would engage in a dialogue if the player

Table 7.6
Research using the media comparison approach in social studies

Source	Game version	Control	Topic	Learners	Effect size
Social studies					
Virvou et al. (2005)	*VR-ENGAGE*	Intelligent tutoring system	Geography	Elementary (low skill)	1.42
Virvou et al. (2005)	*VR-ENGAGE*	Intelligent tutoring system	Geography	Elementary (high skill)	–0.16
Huizenga et al. (2009)	*Frequency 1550*	Project-based lesson	Medieval Amsterdam	Secondary	0.62
Median					0.62

was not sure of the answer. The tutoring system asked the same questions and provided feedback, but without any game context. On a hundred-item posttest on geography, low-academic-performing students scored much higher with the game than the control treatment ($d = 1.42$), but high-academic-performing students did slightly worse with the game than the control treatment ($d = -0.16$).

In the third line of table 7.6, Huizenga, Admiraal, Akkerman, and ten Dam (2009) asked 458 Dutch students in their first year of secondary school to learn about medieval Amsterdam by playing the *Frequency 1550* mobile (location-based) game in groups or engaging in regular project-based lessons. The game group outperformed the conventional group on a posttest with both multiple-choice and open-ended questions ($d = 0.62$). The effects of the game were stronger both for preuniversity students than for prevocational students, and for low-prior-knowledge learners rather than high-prior-knowledge learners.

Does Game Effectiveness Depend on Age Group, Type of Game, or Type of Control Group?

Table 7.7 summarizes a final set of analyses aimed at determining whether game effects are stronger for certain age groups, game types, or control groups. These analyses must be interpreted in light of the fact that studies

Table 7.7
Research using the media comparison approach by age group, game type, and type of control group

Type	Number	Effect size
Age group		
Elementary	12 out of 14	0.34
Secondary	4 out of 6	0.58
College/adult	8 out of 12	0.74
Game type		
Quiz/puzzle	10 out of 12	0.45
Simulation	5 out of 7	0.62
Adventure	9 out of 12	0.72
Control group		
Electronic	6 out of 12	0.12
Paper	6 out of 7	0.53
Classroom	12 out of 13	0.63

varied on many dimensions, and were not controlled to focus solely on age, game type, or type of control group.

As summarized in the top portion of table 7.7, the comparisons were categorized into those involving elementary school children (i.e., grades K–6), secondary school students (i.e., grades 7–12), or college students and adults. The median effect size favoring the game group was strongest for college students and adults ($d = 0.74$ based on twelve comparisons) and weakest for elementary school children ($d = 0.34$ based on twelve comparisons).

As shown in the middle portion of table 7.7, the comparisons were categorized into those involving quiz or puzzle games, simulation games, and adventure games. The median effect size favoring the game group was strongest for adventure games ($d = 0.72$ based on twelve comparisons), and weakest for quiz and puzzle games ($d = 0.45$ based on twelve comparisons).

As summarized in the bottom portion of table 7.7, the comparisons were categorized into those involving control groups that received electronic lessons, paper lessons, or face-to-face classroom lessons (e.g., regular classroom activity). The median effect size favoring the game group was strongest when the control group was classroom based ($d = 0.63$ in thirteen comparisons) and weakest when it was computer based ($d = 0.12$ in twelve comparisons). This finding is somewhat disconcerting because it suggests that when more experimental control is applied (i.e., both groups learn with computer-based platforms), the effectiveness of games appears to be eliminated.

Discussion

Critique of the Media Comparison Approach to Game Research

Do computer games promote learning? In some ways this is not a fruitful question because media comparison studies such as those summarized in this review are highly susceptible to four kinds of serious problems: conceptual, methodological, empirical, and theoretical. On the conceptual side, Clark (2001) has eloquently distinguished between instructional media (such as games versus conventional media) and instructional methods (such as discovery versus direct instruction). According to Clark's analysis, as mentioned earlier, instructional media do not cause learning but instructional methods do cause learning. Studies that confound media and methods are in jeopardy of attributing learning outcome effects to instructional media when, in fact, they are caused by differences in instructional methods. Thus, media comparison studies may be conceptually problematic to the extent that they confound media and method. A solution to this conceptual problem is to conduct value-added research within one medium, such as comparing the learning outcomes of students who receive a base version of a game as opposed to the same game with one instructional feature added (as explored in chapter 5).

On the methodological side, media comparison experiments are particularly at risk for violating the scientific requirement of experimental control, in which the experimental and control groups are identical in all ways except for the dimension that is being varied. When a computer-based game is compared to a face-to-face lesson, for example, there are so many dimensions on which the two treatments differ that it may not be possible to isolate what causes any differences in learning outcome. A related aspect of the experimental control problem is that in some experiments, it is not clear that game and control groups were exposed to the same material for the same amount of time. Equating the game and conventional groups on relevant dimensions is a serious challenge that is hard to overcome in media research.

On the empirical side, media comparison studies are particularly susceptible to the file-drawer problem in which only "successful" studies are published in academic journals (Ellis, 2010). For this reason, it would be useful to determine whether the results of dissertations and conference papers yield as strong effects as do published papers. The same problem, moreover, can occur when researchers use multiple dependent measures of learning outcome and only report on (or emphasize) those that yield significant differences.

On the theoretical side, media comparison researchers have much more work to do in linking empirical findings with underlying theories based on the motivating properties of games. The present findings do not paint a clear picture of how gaming affects student motivation and how motivation affects students' learning outcomes. What is the role of motivational factors in learning with games? Is there a theoretical explanation for why games might be more effective for some academic disciplines or age groups? These are some of the unanswered theoretical questions in the media comparison approach to game research.

Contributions of the Media Comparison Approach to Game Research

The major finding in this review is that the most promising academic domains for educational games are science ($d = 0.69$ based on sixteen comparisons) and second-language learning ($d = 0.96$ based on five comparisons). These areas are ripe for focused testing of which features contribute most strongly to game effectiveness.

A secondary finding is that language arts and social studies do not yet have a sufficient evidence base, so additional well-designed media comparison studies are warranted. The same conclusion applies to mathematics learning, which has been shown to be an unpromising domain for games based on preliminary research ($d = 0.03$ based on five comparisons).

For the conceptual and methodological reasons noted above, it is premature to recommend the incorporation of games in school curricula or as stand-alone training venues. A particularly telling finding is that game effectiveness was found to be strongest when the control group was regular classroom activity ($d = 0.63$ based on thirteen comparisons), which represents the poorest experimental control and was weakest when the control group received computer-based instruction ($d = 0.12$ based on ten comparisons), which represents the best experimental control. In short, the present review helps confirm some of the criticisms of the media comparison approach to game research and points to the need for experimental control in media comparison experiments.

Given the conceptual and methodological dangers of media comparison research, should we stop testing the game superiority hypothesis—that is, the idea that people learn better from games than from conventional media? The media comparison approach to game research is a response to the plethora of claims for the game superiority hypothesis. As long as game proponents continue to make this claim, researchers will be inclined to seek scientific evidence to test it. Even if games are ultimately shown to be less effective than conventional media, they may be of educational value in situations where students would choose to spend their free time playing games but would not choose to spend their free time studying with conventional media.

Based on this review, I offer three recommendations concerning the effort to test the game superiority hypothesis. First, when conducting media comparison studies, exercise as much experimental control as possible by striving to ensure that the treatment and conventional groups are as equivalent as possible on all dimensions except the use of gaming. Second, given that the game superiority hypothesis is grounded in motivational theories, also include motivational measures and relate them to learning outcome measures. Third, consider supplementing media comparison studies with experiments that use a value-added approach to game research in which a base version of a game is compared to the same game with one instructional feature added (as described in chapter 5).

References

Note: Asterisk (*) indicates study is in review database.

Abt, C. C. (1970). *Serious games*. New York: Viking.

*Adams, D. M., Mayer, R. E., MacNamara, A., Koening, A., & Wainess, R. (2012). Narrative games for learning: Testing the discovery and narrative hypotheses. *Journal of Educational Psychology, 104*, 235–249.

*Anderson, J., & Barnett, M. (2011). Using video games to support pre-service elementary teachers learning of basic physics principles. *Journal of Science Education and Technology, 20*, 347–362.

Anderson, L. W., Krathwohl, D. R., Airasian, P. W., Cruikshank, K. A., Mayer, R. E., Pintrich, P. R., et al. (2001). *A taxonomy for learning, teaching, and assessing*. New York: Longman.

*Barab, S. A., Scott, B., Siyahhan, S., Goldstone, R., Ingram-Goble, A., Zuiker, S. J., et al. (2009). Transformational play as a curricular scaffold: Using videogames to support science education. *Journal of Science Education and Technology, 18*, 305–320.

*Brom, C., Preuss, M., & Klement, D. (2011). Are educational computer micro-games engaging and effective for knowledge acquisition at high schools? A quasi-experimental study. *Computers & Education, 57*, 1971–1988.

*Chang, K.-E., Wu, L.-J., Weng, S.-E., & Sung, Y.-T. (2012). Embedding game-based problem-solving phase into problem-posing system for mathematics learning. *Computers & Education, 58*, 775–786.

Clark, R. E. (2001). *Learning from media*. Greenwich, CT: Information Age Publishing.

Clark, R. E., Yates, K., Early, S., & Moulton, K. (2011). An analysis of the failure of electronic media and discovery-based learning: Evidence for the performance benefits of guided learning methods. In K. H. Silber & W. R. Foshay (Eds.), *Handbook of improving performance in the workplace* (pp. 263–297). San Francisco: Pfeiffer.

Cohen, J. (1988). *Statistical power analysis for the behavioral sciences* (2nd ed.). Hillsdale, NJ: Erlbaum.

Connolly, T. M., Boyle, E. A., MacArthur, E., Hainey, T., & Boyle, J. M. (2012). A systematic review of empirical evidence on computer games and serious games. *Computers & Education, 59*, 661–686.

*Din, F. S., & Calao, J. (2001). The effects of playing educational video games on kindergarten achievement. *Child Study Journal, 31*, 95–102.

Ellis, P. D. (2010). *The essential guide to effect sizes*. New York: Cambridge University Press.

*Evans, K. L., Yaron, D., & Leinhardt, G. (2008). Learning stoichiometry: A comparison of text and multimedia formats. *Chemistry Education Research and Practice, 9*, 208–218.

Gee, J. P. (2007). *Good video games and good learning*. New York: Peter Lang.

Hannafin, R. D., & Vermillion, J. R. (2008). Technology in the classroom. In T. L. Good (Ed.), *21st century education: A reference handbook* (Vol. 2; pp. 209–218). Thousand Oaks, CA: Sage.

Hayes, R. T. (2005). *The effectiveness of instructional games: A literature review and discussion*. (Technical report 2005–004). Orlando, FL: Naval Air Warfare Center Training Systems Division.

*Hickey, D. T., Ingram-Goble, A. A., & Jameson, E. M. (2009). Designing assessments and assign designs in virtual educational environments. *Journal of Science Education and Technology, 18*, 187–208.

Honey, M., & Hilton, M. (Eds.). (2011). *Learning science through computer games and simulations*. Washington, DC: National Academy Press.

*Huizenga, J., Admiraal, W., Akkerman, S., & ten Dam, G. (2009). Mobile game-based learning in secondary education: Engagement, motivation, and learning in a mobile city game. *Journal of Computer Assisted Learning, 25*, 332–344.

*Hwang, G.-C., Wu, P.-H., & Chen, C.-C. (2012). An online game approach for improving students' learning performance in Web-based problem-solving activities. *Computers & Education, 59*, 1246–1256.

Katona, G. (1940). *Organizing and memorizing*. New York: Columbia University Press.

Lepper, M. R., & Malone, T. W. (1987). Intrinsic motivation and instructional effectiveness in computer-based education. In R. E. Snow & M. J. Farr (Eds.), *Aptitude, learning, and instruction: III. Conative and affective process analysis* (pp. 255–286). Hillsdale, NJ: Erlbaum.

*Liu, T., & Chu, Y. (2008). Using ubiquitous games in an English listening and speaking course: Impact on learning outcomes and motivation. *Computers & Education, 55*, 630–643.

Malone, T. W., & Lepper, M. R. (1987). Making learning fun: A taxonomy of intrinsic motivation for learning. In R. E. Snow & M. J. Farr (Eds.), *Aptitude, learning, and instruction: III. Conative and affective process analysis* (pp. 223–253). Hillsdale, NJ: Erlbaum.

Mayer, R. E. (2009). *Multimedia learning* (2nd ed.). New York: Cambridge University Press.

Mayer, R. E. (2011). Multimedia learning and games. In S. Tobias & J. D. Fletcher (Eds.), *Computer games and instruction* (pp. 281–306). Amsterdam: Elsevier.

*Moreno, R., Mayer, R. E., Spires, H., & Lester, J. C. (2001). The case for social agency in computer-based teaching: Do students learn more deeply when they interact with animated pedagogical agents? *Cognition and Instruction, 19*, 177–213.

Mosteller, F., & Boruch, R. (Eds.). (2002). *Evidence matters: Randomized trails in education research.* Washington, DC: Brookings Institution Press.

*Neri, A., Mich, O., Gerosa, M., & Giuliani, D. (2008). The effectiveness of computer-assisted training for foreign language learning by children. *Computer Assisted Language Learning, 21*, 393–408.

O'Neil, H. F., & Perez, R. S. (Eds.). (2008). *Computer games and team and individual learning.* Amsterdam: Elsevier.

O'Neil, H. F., Wainess, R., & Baker, E. L. (2005). Classification of learning outcomes: Evidence from the computer games literature. *Curriculum Journal, 16*, 455–474.

*Parchman, S. W., Ellis, J. A., Christinaz, D., & Vogel, M. (2000). An evaluation of three computer-based instructional strategies in basic electricity and electronics training. *Military Psychology, 12*, 73–87.

*Papastergiou, M. (2009). Digital game-based learning in high school computer science education: Impact on educational effectiveness and student motivation. *Computers & Education, 52*, 1–12.

Phye, G. D., Robinson, D. H., & Levin, J. (Eds.). (2005). *Empirical methods for evaluating educational interventions.* San Diego: Academic Press.

Raessens, J., & Goldstein, J. (Eds.). (2005). *Handbook of computer game studies.* Cambridge, MA: MIT Press.

Randel, J. M., Morris, B. A., Wetzel, C. D., & Whitehill, B. V. (1992). The effectiveness of games for educational purposes: A review of recent research. *Simulation & Games, 23*, 261–276.

*Ricci, K. E., Salas, E., & Cannon-Bowers, J. A. (1996). Do computer-based games facilitate knowledge acquisition and retention? *Military Psychology, 8*, 295–307.

Rosenthal, R., Rosnow, R. L., & Rubin, D. B. (2000). *Contrasts and effect sizes in behavioral research*. New York: Cambridge University Press.

Schneider, B., Carnoy, M., Kilpatrick, J., Schmidt, W. H., & Shavelson, R. J. (2007). *Estimating causal effects using experimental and observational designs*. Washington, DC: American Educational Research Association.

*Segers, E., & Verhoeven, L. (2003). Effects of vocabulary training by computer in kindergarten. *Journal of Computer Assisted Learning, 2003*, 557–566.

Shaffer, D. W. (2006). *How computer games help children learn*. New York: Palgrave Macmillan.

Shavelson, R. J., & Towne, L. (Eds.). (2002). *Scientific research in education*. Washington, DC: National Academy Press.

*Sindre, G., Natvig, L., & Jahre, M. (2009). Experimental validation of the learning effect for a pedagogical game on computer fundamentals. *IEEE Transactions on Education, 52*, 10–18.

*Suh, S., Kim, S. W., & Kim, N. J. (2010). Effectiveness of MMORPG-based instruction in elementary English education in Korea. *Journal of Computer Assisted Learning, 26*, 370–378.

*Swaak, J., de Jong, T., & van Joolingen, W. R. (2004). The effects of discovery learning and expository instruction on the acquisition of definitional and intuitive knowledge. *Journal of Computer Assisted Learning, 20*, 225–234.

Tobias, S., & Fletcher, J. D. (Eds.). (2011). *Computer games and instruction*. Charlotte, NC: Information Age Publishers.

Tobias, S., Fletcher, J. D., Dai, D. Y., & Wind, A. P. (2011). Review of research on computer games. In S. Tobias & J. D. Fletcher (Eds.), *Computer games and instruction* (pp. 525–545). Charlotte, NC: Information Age Publishing.

Van Eck, R. (Ed.). (2010). *Gaming and cognition: Theories and practice from the learning sciences*. Hershey, PA: Information Science Reference.

*Van Eck, R., & Dempsey, J. (2003). The effect of competition and contextualized advisement on the transfer of mathematics skills in a computer-based instructional simulation game. *Educational Technology Research and Development, 50*, 23–41.

*Virvou, M., Katsionis, G., & Manos, K. (2005). Combining software games with education: Evaluation of its educational effectiveness. *Journal of Educational Technology & Society, 8*(2), 54–65.

Vogel, J. F., Vogel, D. S., Cannon-Bowers, J., Bowers, C. A., Muse, K., & Wright, M. (2006). Computer gaming and interactive simulations for learning: A meta-analysis. *Journal of Educational Computing Research, 34*, 229–243.

Vorderer, P., & Bryant, J. (Eds.). (2006). *Playing video games: Motives, responses, and consequences*. Mahwah, NJ: Erlbaum.

*Wrzesien, M., & Raya, M. A. (2010). Learning in serious virtual worlds: Evaluation of learning effectiveness and appeal to students in the E-Junior project. *Computers & Education, 55*, 178–187.

*Yip, F.W.M., & Kwan, A.C.M. (2006). Online vocabulary games as a tool for teaching and learning English vocabulary. *Educational Media International, 43*, 233–249.

Young, M. F., Slota, S., Cutter, A. B., Jalette, G., Mullin, G., Lai, B., et al. (2012). Our princess is in another castle: A review of trends in serious gaming for education. *Review of Educational Research, 82*, 61–89.

III Conclusion

8 The Future of Research on Games for Learning

Chapter Outline

Summary

The overall theme of this book is that the design of games for learning should be based on rigorous scientific evidence rather than on strong opinions. The value-added approach is effective when the research question is, Which features improve a game's effectiveness? The cognitive consequences approach is effective when the research question is, What is learned from playing a game? The media comparison approach faces some serious conceptual and methodological challenges in equating the academic content of the game group and the conventional group, but it can provide useful evidence when the research question is, Are games more effective than conventional media? In addition to collecting replicated evidence on these three questions, the next steps include continuing to systemize the available research (as is exemplified in chapters 4 through 7), identifying boundary conditions under which design principles are most likely to be effective, and developing a theory of learning with games that includes testable mechanisms of learning and motivation. Experimental methods are most appropriate for the three research questions addressed in this book, although they should be tempered with appropriate instrumentation for measuring learning outcomes and processes. Finally, three nonrecommended paths for the future are advocacy research (which is based on anecdotes, testimonials, and flawed studies), anything-but-learning research (which fails to measure learning outcomes), and let's-see-what-happens research (which can easily overwhelm us with detailed but useless descriptions of game-playing activity).

Research Summary: Where We Are Coming From

The Struggle for an Evidence-Based Approach

Many strong claims are made for the potential of games to revolutionize education, but such claims are rarely backed up with convincing scientific evidence. Consider the following three steps in the claims-based approach to educational computer games.

Step 1 Observe external behavior. We begin our journey with the observation that computer games are popular. There is ample evidence that people spend a lot of their free time engaged in playing computer games and spend a lot of money to do so (Kapp, 2012).

Step 2 Infer internal mechanisms. The next step is to infer that people like playing games, or in more academic jargon, game playing is motivating for people. There is some anecdotal and observational evidence that people initiate game play, persist in game play, and engage intensely in game play—all of which are indexes of motivation.

Step 3 Call for revolutionizing education. As lack of motivation is a crucial problem for formal education, visionaries call for revolutionizing schooling by using what we know about the motivating power of computer games.

What is wrong with this three-step journey to the gamification of education? The problem with this argument is that it based on strong claims rather than scientific evidence. Just because game playing is a powerful source of entertainment does not automatically mean it is a powerful tool for education.

Consider a similar argument that was made nearly a hundred years ago for what was then a new technology for entertainment: motion pictures. In 1922, Thomas Edison had high hopes for the educational potential of the new entertainment medium of motion pictures: "I believe that the motion picture is destined to revolutionize our educational system and that in a few years it will supplant largely, if not entirely, the use of textbooks" (quoted in Cuban, 1986, p. 9). Like computer games, motion pictures are a popular form of entertainment, but over their hundred-year history they have not yet played the kind of revolutionary role in education that was predicted. Similar claims were made in the 1930s and 1940s for how radio would revolutionize education, and in the 1950s for how television would revolutionize education, yet these entertainment media failed to live up to the expectations (Cuban, 1986). This history of entertainment technology's lack of impact on education is part of the reason to be cautious in assessing parallel claims made today for how games are destined to revolutionize education.

The main theme of this book is that although grand claims for the educational potential of educational games may attract our attention, an important step is to conduct scientific research that will generate needed evidence concerning the effectiveness of computer games in education. The value of an evidence-based approach is that it is self-correcting in a way that an opinion-based approach is not—that is, evidence is helpful in distinguishing between what actually works in game design versus what proponents say will work. In taking an evidence-based approach, we begin with testable research questions such as the value-added, cognitive consequences, and media comparison questions (as described in chapter 1); we seek appropriate research methods to address the questions such as the experimental comparisons used in the value-added, cognitive consequences, and media comparison approaches (as described in chapter 2); we seek to ground the research in cognitive theories of how people learn (as described in chapter 3); and we seek to create a database of research evidence that has practical and theoretical implications for game design (as described in chapters 4 through 7).

Contributions of Value-Added Research

Value-added research shows promise for both practice and theory in game design. Concerning practice, value-added research can suggest features that increase the instructional effectiveness of games. As explored in chapter 5, some promising evidence-based design principles are modality (i.e., using spoken words), personalization (i.e., using conversational style), pretraining (i.e., providing pregame experiences for foundational knowledge), coaching (i.e., offering advice or explanations throughout the game), and self-explanation (i.e., asking people to give explanations during the game). In contrast to the support for these features, research does not support the instructional value of immersion (i.e., using realistic 3-D renderings) and redundancy (i.e., presenting both spoken and printed words at the same time), and there is not yet convincing research to support the use of popular features such as competition (e.g., giving prizes based on scores), segmenting (e.g., breaking screens into windows), image (e.g., having a talking head on the screen), narrative theme (e.g., having a detailed story line), choice (e.g. allowing players to choose the look of the screen), and learner control (e.g., allowing learners to control the order and pace of presentation).

Concerning theory, research using the value-added approach contributes to cognitive theories of learning by showing that the motivational features of games can be combined with instructional features to help prime appropriate cognitive processing in learners, as discussed in chapter 3. Overall, the value-added approach has been shown to be an appropriate research method when the research question involves the instructional effectiveness of adding a new feature to a game. There clearly is much work to be done in pinpointing effective instructional features of games, and the value-added approach is a useful tool for the job.

Contributions of Cognitive Consequences Research

Cognitive consequences research also shows promise for contributing to practice and theory in game design. Concerning practice, the cognitive consequences approach is beginning to identify which kinds of off-the-shelf computer games can lead to improvements in which kinds of cognitive skills. Chapter 6 summarizes evidence that playing certain first-person shooter games can improve specific perceptual and spatial skills, whereas playing certain spatial puzzle games can improve specific mental rotation skills. In contrast, the claims for improving reasoning, memory, and motor performance through game playing do not yet have convincing evidence in the research literature.

As far as theory goes, there is some support for the idea that the positive cognitive consequences of playing games are strongest when the cognitive processing required in the game is the same as the cognitive processing required in the tests of cognitive skill. Overall, the cognitive consequences approach has been shown to be an appropriate research method when the research question concerns the cognitive effects of playing off-the-shelf games. Additional work is needed in identifying off-the-shelf games that produce positive cognitive consequences in learners, and the cognitive consequences approach is a useful tool for that job.

Contributions of Media Comparison Research

Finally, media comparison research is the method of choice when the research question concerns whether academic content is best learned by playing a game or through conventional media. Media comparison studies are used to justify the cost and effort of creating educational computer games. Some preliminary implications for practice explored in chapter 7 are that games may be more effective in science and second-language learning than in math, when the learners are college students rather than elementary school children, and when the control group is standard instruction in a classroom instead of a well-controlled electronic medium such as a slideshow.

The theoretical implications are that differences appear to be small when the contents and instructional methods of a game and a conventional medium share the most similarities, thereby suggesting that the instructional medium per se does not cause learning. As discussed in chapter 7, media comparison research has a somewhat-disappointing history in educational technology, owing largely to the huge methodological and conceptual challenges of separating instructional media (e.g., games versus slideshow presentation) from instructional methods (e.g., practice with feedback versus direct instruction). Thus, the media comparison approach should be used with a healthy dose of caution concerning the degree to which the game and conventional groups receive the same content and instructional methods.

Research Agenda: Where We Are Going

Role of Research Questions

Game research can advance to the degree that researchers ask appropriate questions—that is, questions that can be addressed in scientifically rigorous research, and that have useful theoretical and practical implications. A primary theme of this book is that researchers have been asking three basic questions about educational games: the value-added question, Which features improve a game's effectiveness?; the cognitive consequences question, What is learned from playing a game?; and the media comparison question, Are games more effective than conventional media? In my opinion, these can be productive questions so it makes sense to continue with them. Intensive research on these questions has begun, but is still in its early stage. This book provides you with a look at what the research has to say about these questions up to this point, which I call the *initial explorations phase*.

As a next step, it is worthwhile to continue asking these questions in order to build a solid base of research evidence. The field of game research will advance to the degree that researchers can amass a body of replicated findings that have useful theoretical and practical implications, and can be organized in a systematic way. The key new question is, How do the findings fit together into a coherent picture? A primary goal of this book is to show how the growing research base can be systematized, which I call the *systematic organizing phase*. This book offers you an initial organization by pinpointing features that have and have not been shown to increase the instructional effectiveness of games, identifying the cognitive skills that are susceptible to improvement through game playing, and distinguishing which kinds of game-playing activities are better or worse than learning with conventional media.

As the field progresses, researchers should be asking an important collection of secondary questions to establish the boundary conditions of each key effect: For which kinds of learners is the effect strong? For which kinds of game-playing situations is the effect strong? For which kinds of outcome measure is the effect strong? The goal of research on boundary conditions is to test theory-based predictions for when game effects should be strongest and weakest. In this book, I have highlighted information about boundary conditions when it is available, which I call the *boundary-building phase*.

Researchers ultimately should address questions about the learning and motivational mechanisms underlying strong game effects, such as, Which learning processes are activated? Which motivational processes are activated? How are they coordinated during learning? These questions constitute the basis for what I call the *theory-building phase*.

Finally, a major advancement in game research will occur when we can apply what has been learned to improving education in ways that actually work. The reciprocal questions here are: How can we use game research and research-based theory to improve education? How can we use attempts to improve education to improve game research and research-based theory? This is what I call the *research application phase*, in which game research can have both practical and theoretical goals.

In short, the steps in a research agenda can be defined in terms of the questions asked: initial explorations, systemic organizing, boundary building, theory building, and research application.

Role of Research Methods

Research questions motivate us to get started on scientific research, but research methods are the tools we use to get the job done. As Shavelson and Towne (2002) so eloquently state, educational researchers should "use methods that permit direct investigation of the question" (p. 3). In the case of the three questions about instructional effectiveness at the center of this book, the most appropriate methodology involves the use of experimental comparisons, as described in chapter 2.

Are there viable alternatives to experimental methods in game research? As noted in chapter 2, the research method depends on the question we wish to answer. *Experimental methods* are particularly helpful in addressing questions about the causes of instructional effectiveness, such as whether one version of a game is more effective than another (value-added research), whether playing a game causes changes in cognitive skill as compared to not playing (cognitive consequences research), and whether playing a game is more effective than learning with conventional media (media comparison research). There are certainly other worthwhile questions to ask, of course, but I have focused on these three in order to systematize a growing research base that has potential to guide game design.

Observational methods offer an important set of alternative methodologies, including descriptions of game-playing activity, interviews with game players, and surveys of game players. When the research questions involve how people play games, then observational studies are called for. Observing someone's game playing or asking game players to report on their game playing have a long and useful history in the field, including Turkle's (1995) classic *Life on the Screen*. Descriptions and self-reports provide a richness of detail that is missing in some experimental research. I have chosen not to concentrate on these kinds of questions in this book in order to look more clearly at basic questions about instructional effectiveness.

Development methods focus on creating new learning technologies and informally testing them with users. For example, many of the science games and simulations reviewed in Honey and Hilton's (2011) *Learning Science through Computer Games and Simulations* are based on development projects in which researchers create exciting computer-based simulation games and informally observe how students use them. Documenting the history of a game's development can be a useful contribution, but it does not directly address the questions posed in this book.

Finally, nonscientific approaches are those that substitute anecdotes, testimonials, and grand visions of the future for solid research evidence. All I can say here is that proselytizing is not a scientific research method.

Overall, experimental methods are best suited for answering the three kinds of questions posed in this book. Improvements are needed in assessing learning outcomes, including making better use of embedded assessment; assessing cognitive processing, cognitive load, and motivational processes during learning, including making sense of in-game measures and physiological measures; and pinpointing the role of individual difference in learners, the conditions of game playing, longitudinal changes over time in playing games, and how games are integrated into an educational context.

Role of Research Theories

Current theories of learning with games are somewhat underdeveloped, especially concerning the interaction of cognitive and motivational processes in learning. The emerging research base on games for learning should help to better delineate the mechanisms of learning and motivation, including the underlying cognitive and motivational processes as well as how they are controlled. A major theoretical contribution of game research is to integrate both cognitive and motivational processes in a theory of learning with games. As theories become better specified, they can suggest specific research predictions for empirical testing.

Role of Research Instrumentation

The measurement of learning outcomes is a crucial aspect of research on games for learning. Although in the research reviewed in this book assessment is commonly accomplished through a separate paper-and-pencil test given after game playing, an exciting next step is the development of embedded tests, which assess learning outcomes within the context of playing the game. The assessment of cognitive and motivational processes during learning can be advanced through the clever use of biophysical technologies such as eye tracking, recording of facial expression, EEG patterns, fMRI data, heart rate, blood pressure, and so on. Research also is needed to investigate the role of handheld mobile technologies such as tablets and smartphones, or embodied game platforms like the Wii or Kinect. In short, technology can serve to allow more effective measurements of learning as well as affording new methods of instruction with games.

Research Detours: Where Not to Go

Finally, what kind of research is not needed? In my opinion, the three most troubling genres of game research are what I call *advocacy research* (opinion-based scholarship characterized by strong claims based on weak evidence), *anything-but-learning research* (research that focuses on issues such as how much players like the game rather than what they learn from it), and *let's-see-what-happens research* (research aimed at mining activity logs or player video with no theory to guide the analysis).

Advocacy research is reflected in papers and books in which game proponents extol the potential of games to transform education, but do not offer adequate empirical evidence. The hallmark of advocacy research is strong claims coupled with weak evidence. The evidence—if there is any evidence at all—generally involves carefully selected snippets—descriptions of players in the act of playing, intended to demonstrate the learning power of game playing; best practices—descriptions of game environments that are so compelling on the surface that only a stickler would ask, "Do they work?"; testimonials—in which students or teachers report on the benefits of game playing; and methodologically flawed studies—such as research showing a pretest-to-posttest gain for game players, yet lacking a comparison group. When the paper you are reading consists mainly of expert opinion rather that scientifically rigorous evidence, you have located advocacy research. Although this genre of game research may be prominent in today's game literature, in my view it is not likely to advance our field in the future. If you are interested in taking a scientific approach to game research, you will need a strong advocacy research filter.

Anything-but-learning research is reflected in studies in which the major dependent measure is not a measure of what was learned. Instead, a common focus is on how much the game player likes the game—a special case that I call *liking-not-learning research.* Sometimes the main dependent measure is some self-report such as the player's willingness to play again or amount of mental effort, or sometimes the dependent measure may be the amount of aggression displayed in a new situation. The hallmark of anything-but-learning research is that the dependent measures tap anything and everything other than the main goal in most educational studies: a change in the learner's knowledge or skill. Although some nonlearning measures are important for testing educational theory (such as motivational measures of self-efficacy or goal orientation, as discussed in chapter 3), they should be used in conjunction with learning outcome measures and not instead of them.

Let's-see-what-happens research is reflected in observational studies that focus on describing in minute detail what a game player does. When the main data source is hundreds of pages of log files, detailing the game player's every move, or hours of video files, you are likely to be dealing with let's-see-what happens research. I call it *let's see what happens* because the researcher does not have a theory to test but rather is collecting data just to see what can be observed. Certainly, case studies can serve a useful role, as they provide a level of richness that can complement other evidence. However, in my opinion, it is extremely difficult to do this kind of research in a systematic and focused way, and as a result, much let's-see-what-happens research does not contribute much to the field.

In summary, what is not needed is research that takes a nonscientific approach (e.g., advocacy research), research that uses inappropriate dependent measures (e.g., anything-but-learning research), or research that collects data without a use for it (e.g., let's-see-what-happens research). In contrast, this book's theme is that what is needed in game research is scientifically rigorous experiments within the value-added, cognitive consequences, or media comparison approaches. I will consider this book to be a success to the extent that it motivates high-quality research on games for learning that improves educational practice and learning theory.

References

Cuban, L. (1986). *Teachers and machines: The classroom use of technology since 1920.* New York: Teachers College Press.

Honey, M. A., & Hilton, M. (Eds.). (2011). *Learning science through computer games and simulations.* Washington, DC: National Academy Press.

Kapp, K. M. (2012). *The gamification of learning and instruction.* San Francisco: Pfieffer.

Shavelson, R. J., & Towne, L. (Eds.). (2002). *Scientific research in education.* Washington, DC: National Academy Press.

Turkle, S. (1995). *Life on the screen.* New York: Simon and Schuster.

About the Author

Richard E. Mayer is Professor of Psychology at the University of California, Santa Barbara (UCSB) where he has served since 1975. He received a Ph.D. in Psychology from the University of Michigan, and was Visiting Assistant Professor of Psychology at Indiana University for two years. His current research involves the intersection of cognition, instruction, and technology with a special focus on multimedia learning, computer-supported learning, and learning with games. He is past President of Division 15 (Educational Psychology) of the American Psychological Association, past Vice President of the American Educational Research Association for Division C (Learning and Instruction), former editor of the *Educational Psychologist*, former coeditor of *Instructional Science*, and former Chair of the UCSB Department of Psychology. He received of the E. L. Thorndike Award for career achievement in educational psychology, the Sylvia Scribner Award for research contributions in learning and instruction, and the Distinguished Contribution of Applications of Psychology to Education and Training Award from the American Psychological Association. He is ranked #1 as the most productive educational psychologist in the world by *Contemporary Educational Psychology*. He currently is on the editorial boards of 12 journals mainly in educational psychology. He has served as an elected local school board member in Goleta, California since 1981. He is the author of more than 400 publications including 25 books, such as *The Cambridge Handbook of Multimedia Learning: Second Edition* (editor, 2014), *Applying the Science of Learning* (2011), *Handbook of Research on Learning and Instruction* (editor, with P. Alexander, 2011), *e-Learning and the Science of Instruction: Third Edition* (with R. Clark, 2011), *Multimedia Learning: Second Edition* (2009), and *Learning and Instruction: Second Edition* (2008).

Name Index

Note: Page numbers followed by "t" indicate tables.

Subject Index

Note: Page numbers followed by "f" or "t" indicate figures and tables, respectively.